Modulation

Second Edition

F R Connor

Ph D, M Sc, B Sc (Eng) Hons, ACGI,
C Eng, MIEE, MIERE, M Inst P

Edward Arnold
A division of Hodder & Stoughton
LONDON MELBOURNE AUCKLAND

First published in Great Britain 1982
Reprinted 1975, 1977, 1979 (twice), 1981
Second edition 1982
Reprinted 1984, 1987, 1988, 1990, 1992

British Library Cataloguing in Publication Data

Connor, F. R.
Modulation. — 2nd ed.
1. Telecommunication 2. Modulation (Electronics)
I. Title TK5102.5
621.38′043

ISBN 0—7131—3457—7

Printed and bound in Great Britain for Edward Arnold, a division of Hodder and Stoughton Limited, Mill Road, Dunton Green, Sevenoaks, Kent TN13 2YA by Athenaeum Press Ltd, Newcastle upon Tyne.

Preface

In this new edition, various parts of the text have been revised or extended. In particular, a section on FM stereo is included in Chapter 3 and pulse modulation has been revised in Chapter 5 to include more details on PCM noise and a description of DPCM for picture coding. A further treatment of digital modulation and a new section on spread-spectrum modulation conclude Chapter 5. In Chapter 6 on demodulation, a comparative study of detectors has been made and a section on feedback demodulators for FM threshold reception has been included. The chapter concludes with a section on tracking loops and a useful treatment of digital demodulation. The main text has been extended further by a greater use of appendices to cover such new topics as the Hilbert transform, radio receivers, and feedback loops. It is intended for the reader who wishes to pursue the matter further. As in the first edition, numerous worked examples and a number of problems with answers have been included, together with a greater number of useful references for further reading.

The aim of the book is the same as in the first edition, with the exception that Higher National Certificates and Higher National Diplomas are being superseded by Higher Certificates and Higher Diplomas of the Technician Education Council.

In conclusion, the author would like to express his gratitude to those of his readers who so kindly sent in various comments and corrections for the earlier edition.

1982 FRC

Preface to the first edition

This is an introductory book on the important topic of *Modulation*. The transmission of information requires some form of modulation and the book provides a broad outline of the most important methods used in practice. Moreover, to assist in the assimilation of basic ideas, many worked examples from past examination papers are provided to illustrate clearly the application of the fundamental theory.

The first part of the book is devoted mainly to analogue methods employed in present day systems, such as amplitude modulation and frequency modulation. Some consideration is then given to phase modulation and the various types of pulse modulation in current use, such as pulse code modulation. The book ends with a chapter devoted to the alternative problem of demodulation at the receiver, and the treatment covers the important methods used at present.

This book will be found useful by students preparing for London University examinations, degrees of the Council of Academic Awards, examinations of the Council of Engineering Institutions, and for other qualifications such as Higher National Certificates, Higher National Diplomas, and certain examinations of the City and Guilds of London Institute. It will also be useful to practising engineers in industry who require a ready source of basic knowledge to help them in their applied work.

1973 FRC

Acknowledgements

The author sincerely wishes to thank the Senate of the University of London and the Council of Engineering Institutions for permission to include questions from past examination papers. The solutions and answers provided are his own and he accepts full responsibility for them.

Sincere thanks are also due to the City and Guilds of London Institute for permission to include questions from past examination papers. The Institute is in no way responsible for the solutions and answers provided.

Finally, the author wishes to thank the publishers for various useful suggestions and will be grateful to his readers for drawing his attention to any errors which may have occurred.

Contents

Symbols

f	frequency
f_d	difference frequency
f_i	instantaneous frequency
f_h	highest modulating frequency
f_n	frequency of noise component
f_s	sampling frequency
g_m	mutual conductance
k	any constant
m	modulation factor (depth of modulation)
m_f	frequency modulation index
m_p	phase modulation index
n	any number
A	peak amplitude
B	baseband bandwidth
B_c	channel bandwidth
C	capacitance
	communication capacity
C_j	junction capacitance
E	bit energy
$G(f)$	amplitude spectral density
$J(n)$	Bessel function of order n
L	inductance
N	average noise power
N_o	noise power spectral density
P_c	average carrier power
P_e	probability of error
P_i	input power
P_o	output power
R	resistance
$R(\tau)$	autocorrelation function
$S(f)$	power spectral density
S/N	signal-to-noise ratio
S_i/N_i	input signal-to-noise ratio
S_o/N_o	output signal-to-noise ratio
T	sampling period
W	system bandwidth
	highest frequency component
δ	deviation ratio
Δf_c	peak frequency deviation of a carrier wave
$\Delta \phi$	peak phase deviation of a carrier wave
Δt	peak time displacement
μ	variable parameter

ρ	correlation coefficient
σ	rms noise voltage
	voltage spacing between quantised levels
τ	pulse width
ϕ_i	instantaneous phase angle
ω_c	angular carrier frequency
ω_m	angular modulating frequency
ω_s	angular sampling frequency

Abbreviations

C.E.I.	Council of Engineering Institutions examination in Communication Engineering, Part 2
C.G.L.I.	City and Guilds of London Institute examinations
U.L.	University of London, B Sc (Eng) examination in Telecommunication, Part 3

1
Introduction

In order to transmit information, a form of modulation is necessary. The process of modulation is to vary some parameter of a basic electromagnetic wave usually called the *carrier wave*. A radio-frequency signal is normally used, as the original information is in a form which is unsuitable for distant transmission directly and it is necessary to convert it to a higher frequency by the process of modulation.

Over the years, various modulating methods have been devised for transmitting the required information as effectively as possible with the minimum amount of distortion. The primary factors to be considered are signal power, bandwidth, distortion, and noise power. Ultimately, it is the ratio of signal power to noise power or output *signal-to-noise* ratio (S/N ratio) specified for the system which determines its performance.

Consequently, it is not surprising to find a wide range of modulating techniques being used which appear to compete with one another under given practical conditions. Broadly speaking, these various techniques may be grouped into *analogue* methods which use a sine wave as the carrier signal and *pulse* methods which use a digital or pulse train as the carrier signal. In the past, analogue methods have been very largely exploited, and they are still used because of the capital investment in existing systems and their basic simplicity. More recently, however, other overriding factors have made necessary the use of a pulse modulation technique and so there is a growing use and demand for such methods.

1.1 Analogue methods[1]

The two most important analogue methods are amplitude modulation and angle modulation. Amplitude modulation (AM) with both sidebands and carrier present is most common for certain applications, such as radio broadcasting and radio telephony. The typical AM carrier wave is shown in Fig. 1.1.

More economical versions of AM are vestigial sideband transmission (VSB) which is used in television for economising bandwidth, while double-sideband suppressed carrier (DSBSC) or single-sideband suppressed carrier (SSBSC) provide further power or bandwidth economy. In particular, SSBSC is used

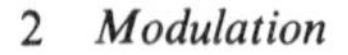

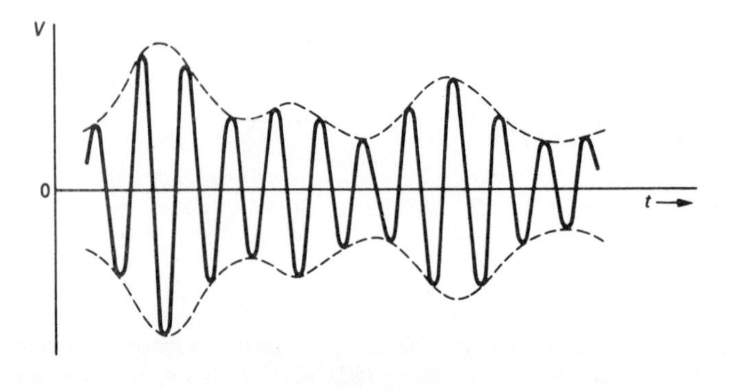

Fig. 1.1

extensively in multiplex coaxial systems for carrying several messages simultaneously. However, the AM systems are essentially narrowband systems and suffer from limitations due to noise which has a direct effect on signal amplitude.

In competition with amplitude modulation, some systems use angle modulation because of its immunity to amplitude-varying noise. In angle modulation, the instantaneous angle of the carrier wave is varied and this leads to two forms of modulation known respectively as *frequency* modulation (FM) and *phase* modulation (PM). As a consequence, FM and PM are closely related though practical systems tend to favour FM rather than PM. Typical examples are VHF broadcasting, satellite communications, and FM radar. However, because the FM carrier wave shown in Fig. 1.2 requires a much greater bandwidth than does its AM counterpart, an FM system is capable of giving a much improved signal-to-noise performance compared to that of the corresponding AM system or, alternatively, a considerable economy in power if required. Hence, FM systems are to some extent superseding AM systems.

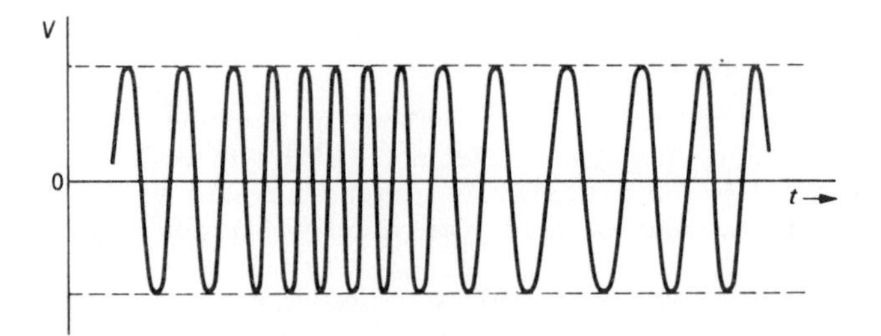

Fig. 1.2

1.2 Pulse methods[2]

An alternative method of modulation uses a digital carrier signal comprising a *pulse train*, which can be modulated to carry the required information. This may be achieved by using the well-established analogue technique of amplitude modulation whereby the pulse amplitude is varied which is called pulse amplitude modulation (PAM). Another technique used is to vary the pulse duration as in pulse duration modulation (PDM) or the time position of the pulses as in pulse position modulation (PPM). The various forms are shown in Fig. 1.3 and it will be shown in Chapter 5 that PAM is linked to amplitude modulation, while PDM and PPM are related to phase modulation. Hence, the signal-to-noise performance improves from PAM through PDM to PPM.

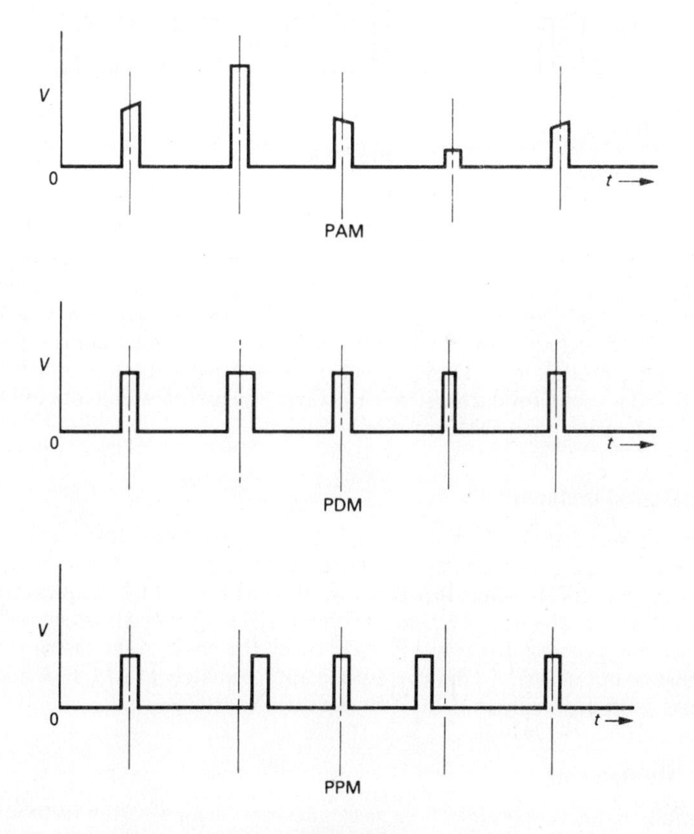

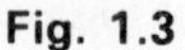
Fig. 1.3

Another form of pulse modulation which has no analogue counterpart is pulse code modulation (PCM) and it promises to be superior to other modulation methods in certain applications. In PCM, information is carried by groups of pulses according to a code as shown in Fig. 1.4. This technique requires a very wide bandwidth but it gives a very good signal-to-noise ratio. It is, in fact, a typical example of the exchange of bandwidth for signal-to-noise ratio and comes nearest to the ideal system as predicted by the Hartley–Shannon law of information theory.

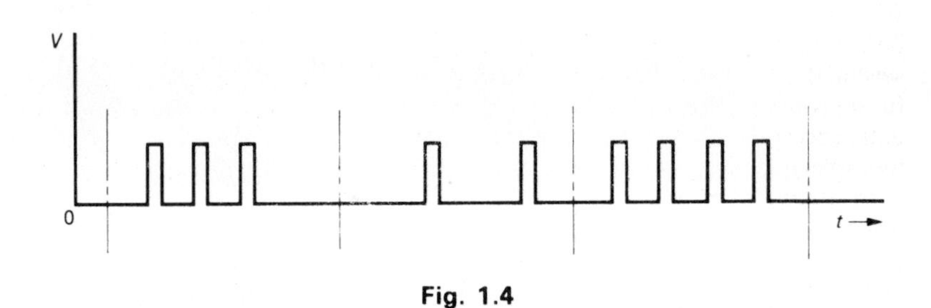

Fig. 1.4

A somewhat simpler form of PCM which is known as delta modulation (DM) has attracted some recent attention. It uses a single pulse code which simplifies equipment design for transmitting and receiving, but is generally found to require a larger bandwidth than PCM. A variation of DM which uses normal PCM coding is known as DPCM. A further development is known as delta sigma modulation (DSM) which is also capable of transmitting d.c. signals and is useful for data systems. However, the present trend is to use PCM or DPCM rather than DM or DSM.

1.3 Digital methods[3]

A form of modulation which uses a digital data signal to modulate a sine-wave carrier is called *digital* modulation. The three types mainly used are amplitude-shift-keying (ASK) in which the carrier wave is either on or off, frequency-shift-keying (FSK) in which the frequency of the carrier is switched from one value to another, and phase-shift-keying (PSK) in which the phase of the carrier is either in phase or out of phase. They correspond approximately to AM, FM, and PM and are used especially in data communication systems.

1.4 Multiplexing

For multichannel communication, analogue or pulse modulation methods may be employed. The multiplexing of analogue signals is well established on a frequency division basis (FDM), especially for use in coaxial cable systems. Increasing demands for more communication channels have led to the development of cables that can now carry up to 10 800 telephone channels each, while microwave links via communication satellites can cope with an even greater number of around 25 000 telephone channels.

However, the multiplexing of channels on a time division basis (TDM) for digital systems is also feasible and appears especially attractive for a PCM system inspite of its sophisticated circuitry. This is because the advent of microelectronics has given PCM a greater impetus and it seems likely that PCM will take over in many short-route systems, such as local telephone circuits, before finding further application in long-distance trunk routes providing several thousand channels at microwave or optical frequencies. The tendency in future communication systems is to transmit all forms of information, such as data, speech, and television, in digital form. More recently, a form of digital multiplexing on a code division basis (CDM)[4] has been investigated using correlation techniques.[5] Multiplexing of the various channels is achieved by using pseudo-random modulation envelopes. At the receiver, a form of correlation detector is required, which has stored information concerning the pseudo-random binary sequences employed as shown in Fig. 1.5.

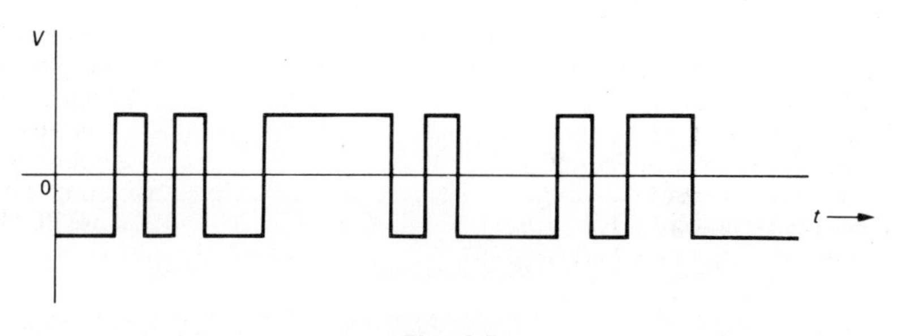

Fig. 1.5

In certain circumstances, this technique offers advantages similar to those of more conventional techniques but, in systems with low signal-to-noise ratio, it is usually inferior and gives worse results than do some of the analogue methods.

2

Amplitude modulation

The process of varying the amplitude of a radio-frequency carrier wave by a modulating voltage is known as amplitude modulation. The carrier amplitude is varied linearly by the modulating signal which usually consists of a range of audio frequencies as for speech or music. To simplify the analysis, consider first the modulating signal as having a single audio frequency and then extend the analysis to the practical case of a more complex audio signal.

Consider now a radio-frequency carrier wave represented by $v_c = V_c \sin \omega_c t$, where $\omega_c = 2\pi f_c$ and f_c is the carrier frequency. If the modulating signal has the form $v_m = V_m \sin \omega_m t$, where $\omega_m = 2\pi f_m$ and f_m is the audio frequency, then it can be seen from Fig. 2.1 that the amplitude of the modulated carrier varies sinusoidally between the values of $(V_c + V_m)$ and $(V_c - V_m)$.

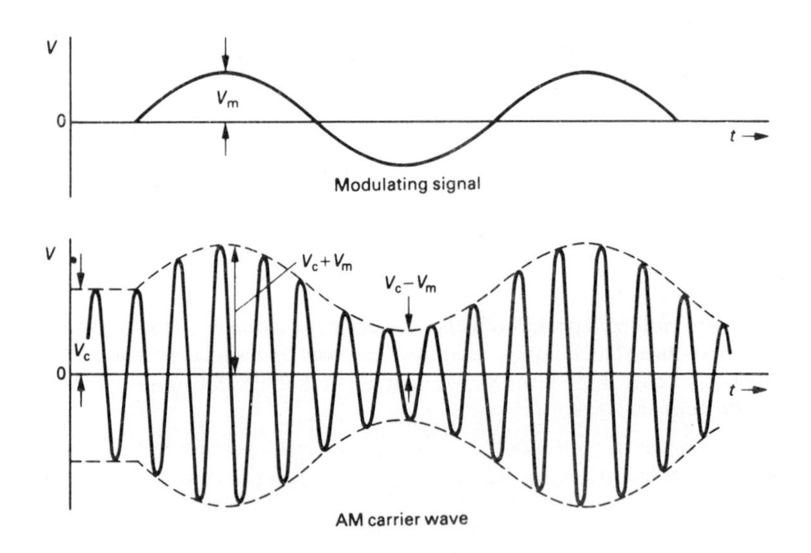

Fig. 2.1

If $V_m/V_c = m$ is the modulation factor (depth of modulation), then $V_m = mV_c$ and the expression for the modulated carrier is

$$v_c = (V_c + V_m \sin \omega_m t) \sin \omega_c t$$

or

$$v_c = V_c \sin \omega_c t + mV_c \sin \omega_c t \sin \omega_m t$$

Now

$$\sin \omega_c t \sin \omega_m t = \tfrac{1}{2}[\cos(\omega_c - \omega_m)t - \cos(\omega_c + \omega_m)t]$$

Hence

$$v_c = V_c \sin \omega_c t + \left(\frac{mV_c}{2}\right)\cos(\omega_c - \omega_m)t - \left(\frac{mV_c}{2}\right)\cos(\omega_c + \omega_m)t$$

2.1 The AM spectrum

The expression above shows that an AM carrier wave contains three frequency components. The frequency of the first term is the carrier frequency, that of the second term is the *lower* side-frequency (LSF), and that of the last term is the *upper* side-frequency (USF). The side-frequencies are above or below the carrier frequency by an amount equal to f_m. Consequently, for a complex modulating signal like speech, numerous frequency components are produced above and below the carrier frequency f_c. They are called the *lower* and *upper* sidebands (LSB and USB) respectively. These are illustrated in Fig. 2.2.

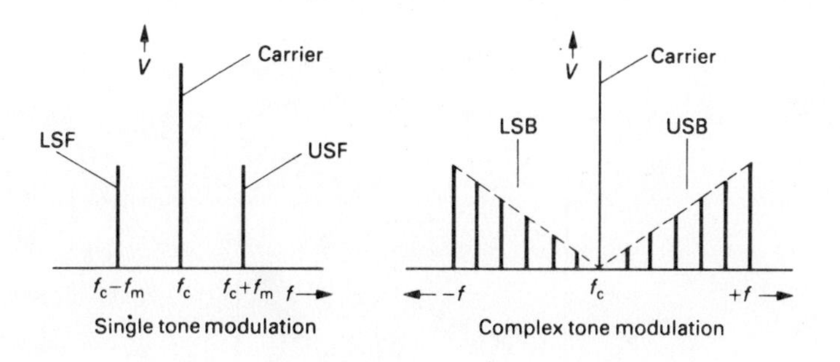

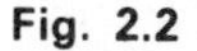
Fig. 2.2

2.2 Power considerations

The average power associated with a signal in a 1 Ω load resistor is equal to its rms voltage squared. For an rms carrier voltage V, the power associated with an AM signal may be expressed as follows

$$\text{carrier power} = V^2$$

$$\text{sideband power} = 2(mV/2)^2 = m^2V^2/2$$

and total power $= V^2 + m^2V^2/2 = V^2(1+m^2/2)$

Hence
$$\frac{\text{sideband power}}{\text{total power}} = \frac{m^2V^2/2}{V^2(1+m^2/2)} = \frac{m^2}{2+m^2}$$

Comments

1. The maximum power in the sidebands is 50% of the carrier power when $m = 1{\cdot}0$.
2. The sideband power depends upon m^2, but for practical reasons m has an average value between 30% and 50%.
3. The carrier and one sideband may be suppressed without destroying the information because the information is also present in the remaining sideband.

2.3 Phasor representation

An a.c. quantity may be represented as a rotating phasor and for the purposes of analysis the instantaneous carrier wave may be regarded as a stationary phasor. The sideband components with frequencies $(f_c + f_m)$ and $(f_c - f_m)$ will then rotate *relative* to the stationary carrier in opposite directions as shown in Fig. 2.3. The resultant voltage is the phasor sum of the three components and represents the AM carrier at any instant of time.

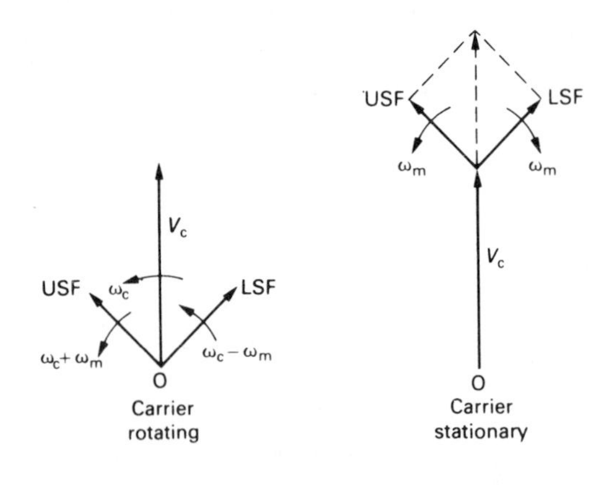

Fig. 2.3

2.4 AM modulators[6]

The *function* of an AM modulator is to modulate a carrier wave, which results in sum and difference frequencies, together with the carrier. This may be achieved by using valves or transistors operating as non-linear or linear modulators.

Non-linear modulators

A non-linear device, such as a semiconductor diode or transistor, may be used. The basic circuit is shown in Fig. 2.4(a) and a practical arrangement in Fig. 2.4(b).

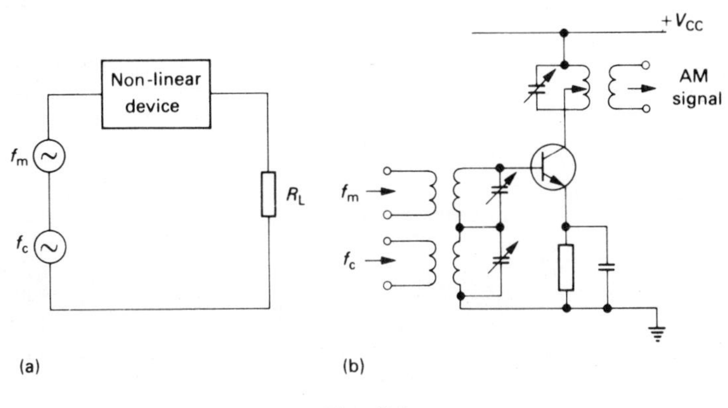

Fig. 2.4

Assuming the device has a characteristic of the form

$$i = a + bv + cv^2$$

where v is the input voltage and i is the output current, for a carrier signal and modulating signal in series at the input, we have

$$v = V_c \sin \omega_c t + V_m \sin \omega_m t$$

$$i = a + b(V_c \sin \omega_c t + V_m \sin \omega_m t) + c(V_c \sin \omega_c t + V_m \sin \omega_m t)^2$$

or

$$i = a + bV_c \sin \omega_c t + bV_m \sin \omega_m t + cV_c^2 \sin^2 \omega_c t + 2cV_c V_m \sin \omega_c t \sin \omega_m t + cV_m^2 \sin^2 \omega_m t$$

Hence

$$i = a + bV_c \sin \omega_c t + bV_m \sin \omega_m t + cV_c V_m \cos(\omega_c - \omega_m)t - cV_c V_m \cos(\omega_c + \omega_m)t + cV_c^2 \sin^2 \omega_c t + cV_m^2 \sin^2 \omega_m t$$

By means of a tuned load in place of R_L, the carrier and sideband components can be selected to give the required AM output. This is shown in Fig. 2.4(b).

Linear modulators

For large power transmitters requiring good linearity, anode modulation is usually employed. The modulator comprises a class-B audio amplifier which drives a class-C RF amplifier and is shown in Fig. 2.5(a).

The class-C amplifier is biased well beyond cut-off and anode current flows in pulses for a part of the RF cycle (Fig. 2.5(b)). The output is rich in harmonics and, by using a tuned anode load, a fairly undistorted AM signal may be obtained with an efficiency of about 80%.

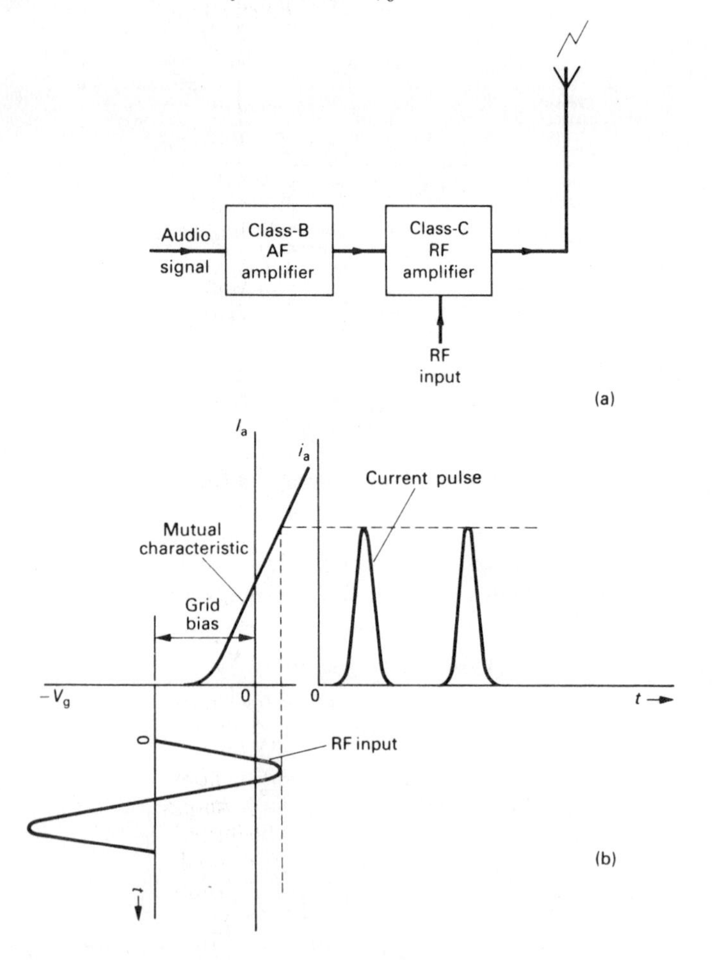

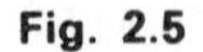
Fig. 2.5

Example 2.1

A push–pull class-B modulator is used to modulate sinusoidally a push–pull class-C RF amplifier. The maximum anode dissipation of the RF amplifier is 250 W and its anode efficiency is 75 %. The class-B modulator has an anode efficiency of 60 % and a maximum anode dissipation of 200 W.

(a) Calculate the maximum modulated RF output from the class-C amplifier.

(b) What is the maximum modulation power that the modulator can supply to the RF amplifier?

(c) What is the maximum depth of modulation? (C.G.L.I.)

Solution

(a) The RF amplifier efficiency η is given by

$$\eta = \frac{\text{output power}}{\text{input power}} = \frac{P_i - P_d}{P_i}$$

where P_d is the anode dissipation. Hence

$$P_i = \frac{P_d}{1-\eta} = \frac{250}{1-0{\cdot}75} = 1000 \text{ W}$$

and the maximum modulated power output P_o becomes

$$P_o = \eta P_i = 0{\cdot}75 \times 1000 = 750 \text{ W}$$

(b) For the modulator we also have

$$\eta = \frac{P_i - P_d}{P_i}$$

where η, P_i, and P_d now refer to the *modulator*. Hence

$$P_i = \frac{P_d}{1-\eta} = \frac{200}{1-0{\cdot}6} = 500 \text{ W}$$

and the maximum modulation power supplied to the RF amplifier is given by

$$P_m = \eta P_i = 0{\cdot}6 \times 500 = 300 \text{ W}$$

(c) Also, $P_o = P_c(1+m^2/2)$, where P_c is the carrier power, $m^2P_c/2$ is the sideband power, and m is the depth of modulation. The sideband power is supplied by the modulator but, since the RF amplifier is only 75 % efficient, we have

$$m^2P_c/2 = 0{\cdot}75 \times P_m = 0{\cdot}75 \times 300 = 225 \text{ W}$$

Hence

$$\frac{m^2P_c/2}{P_c(1+m^2/2)} = \frac{225}{750} = \frac{3}{10}$$

with

$$m^2/(2+m^2) = 3/10$$

and

$$m^2 = 6/7$$

or

$$m = 0{\cdot}926$$

2.5 Other AM systems

(a) DSBSC system

The simplest and most common AM system used is double-sideband transmission in which the carrier and both sets of sidebands are transmitted. Receiver design can be fairly simple and it reduces costs. However, as the carrier does not contain the information, it may be suppressed wholly or partly at the transmitter, yielding a DSBSC system. The saving in carrier power is considerable and amounts to about 96% when $m = 0{\cdot}3$.

Hence, the carrier must be re-inserted at the receiver in order that the original modulation may be recovered. This technique raises complications because of the stringent phase requirements. It is shown in Appendix G that for correct phasing, the modulation is present, but for a 90° phase error, the modulation is lost as no amplitude modulation occurs but only phase modulation. The latter cannot be detected by an AM receiver.

Alternatively, a low-power pilot carrier some 26 dB below its normal value may be transmitted and be used at the receiver to phase-lock a local carrier. Circuits used to generate DSBSC are called *balanced modulators*. A balanced modulator employs non-linear devices, such as valves or transistors. A push–pull arrangement is shown in Fig. 2.6.

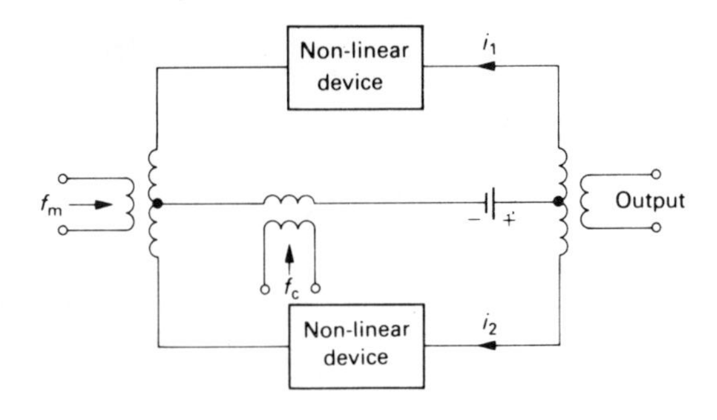

Fig. 2.6

In Fig. 2.6, the carrier is in phase in both halves of the circuit but the modulation is out of phase. The output currents are given by

$$i_1 = k(1 + m \sin \omega_m t) \sin \omega_c t$$
$$i_2 = k(1 - m \sin \omega_m t) \sin \omega_c t$$

where k is a constant of proportionality.

The current flowing in the output transformer is $|i_1 - i_2|$ and we obtain

$$|i_1 - i_2| = 2km \sin \omega_c t \sin \omega_m t$$

or

$$|i_1 - i_2| = km[\cos(\omega_c - \omega_m)t - \cos(\omega_c + \omega_m)t]$$

which contains no carrier component.

An alternative form of balanced modulator uses a bridge arrangement of rectifiers. Typical examples are the Cowan modulator and ring modulator. The rectifiers in each case produce a switching action on the incoming audio signal if the carrier signal is sufficiently large compared to the modulating signal so as to ensure non-linear operation.

Cowan modulator

In Fig. 2.7, when A is positive with respect to B, the diodes conduct and the bridge appears as a short-circuit across the network and no audio signal gets through. When A is negative with respect to B, the diodes are reverse-biased and appear as an open-circuit across the network, thus allowing an audio signal to get through. The carrier wave has a square-wave switching action on the audio signal, which is illustrated in Fig. 2.8.

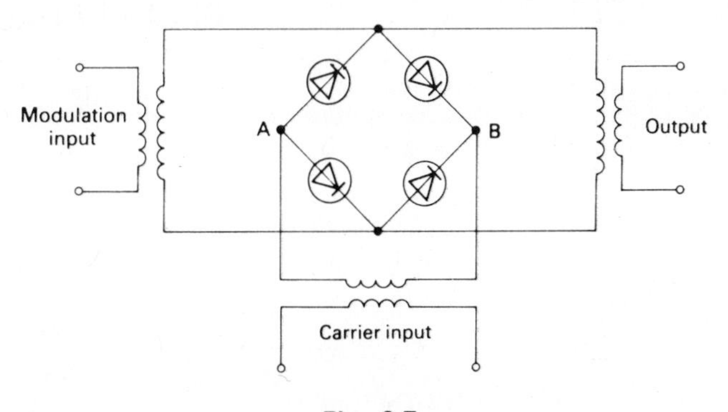

Fig. 2.7

The output signal is the product of the square-wave function and the audio signal. The square-wave switching function is given by

$$v_s = \frac{1}{2} + \frac{2}{\pi}[\sin \omega_c t + \tfrac{1}{3} \sin 3\omega_c t + \ldots]$$

If the audio signal is $v_m = V_m \sin \omega_m t$, then the output v_o is given by

$$v_o = kv_s \times v_m = kV_m \sin \omega_m t \left[\frac{1}{2} + \frac{2}{\pi}\{\sin \omega_c t + \ldots\}\right]$$

or $$v_o = \frac{kV_m \sin \omega_m t}{2} + \frac{kV_m}{\pi}[\cos(\omega_c - \omega_m)t - \cos(\omega_c + \omega_m)t] + \ldots$$

where k is a constant with dimension volt^{-1}.

The output contains the audio signal and the two side-frequencies, but the carrier is absent.

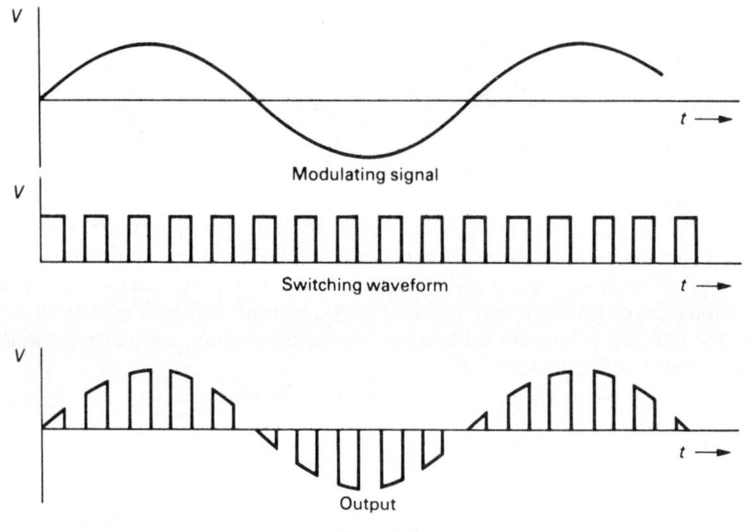

Fig. 2.8

Ring modulator

The circuit is shown in Fig. 2.9 where it will be seen that the carrier is applied to the mid-points of the input and output transformers, while the rectifiers are placed in a 'ring' across the network.

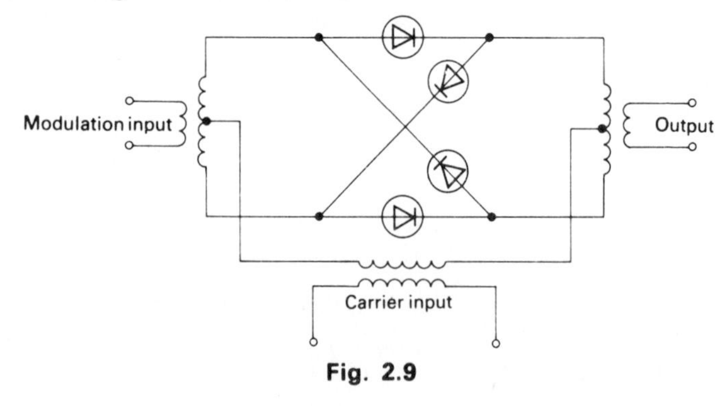

Fig. 2.9

The ring arrangement of rectifiers produces a double-sided square-wave switching action on the network during the positive and negative excursions of the carrier signal. The arrangement acts as a reversing switch operating at the carrier frequency and is illustrated in Fig. 2.10.

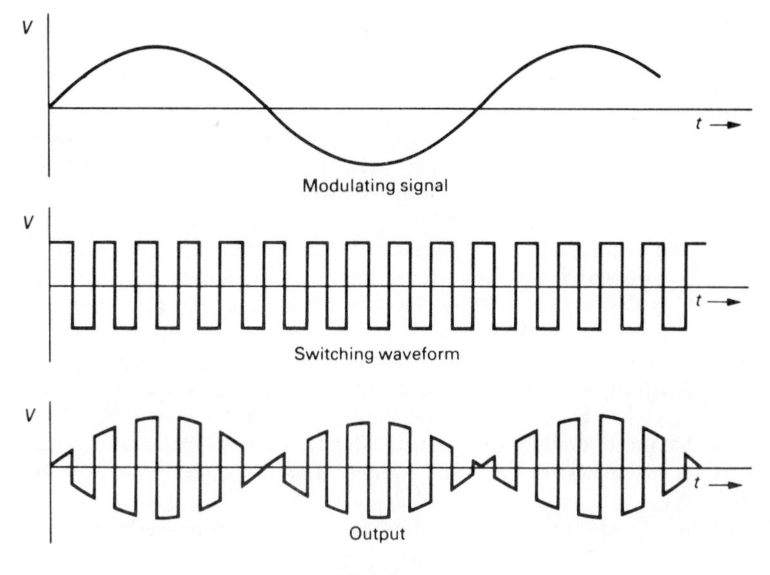

Fig. 2.10

The square wave v_s with a peak amplitude of 1 volt is represented by*

$$v_s = \frac{4}{\pi}[\sin \omega_c t + \tfrac{1}{3}\sin 3\omega_c t + \ldots]$$

If the audio signal is $v_m = V_m \sin \omega_m t$, the output v_o is given by

$$v_o = kv_s \times v_m = kV_m \sin \omega_m t \times \frac{4}{\pi}[\sin \omega_c t + \tfrac{1}{3}\sin 3\omega_c t + \ldots]$$

or

$$v_o = 2\frac{kV_m}{\pi}[\cos(\omega_c - \omega_m)t - \cos(\omega_c + \omega_m)t + \ldots]$$

where k is a constant with dimension volt^{-1}.

The output contains mainly the pair of side-frequencies, but no carrier or audio frequencies are present.

(b) SSBSC system[7,8]

Since the information is present in either sideband, a *further* saving in power and bandwidth may be obtained by suppressing either the lower or the upper

* See F. R. Connor, *Signals*, Edward Arnold (1982).

sideband. The saving in power corresponding to one sideband is approximately 2% of the carrier power for $m = 0{\cdot}3$, while the saving in bandwidth is 50%, since the single sideband covers one half the spectrum of the AM system.

One way of producing an SSBSC signal is to obtain a DSBSC signal by using a balanced modulator and then to suppress the unwanted sideband by a sideband filter. The filter output contains only one sideband as shown in Fig. 2.11.

In practice, difficulties arise in designing a filter with sharp cut-off on either side. Using a narrower bandwidth to give a better cut-off would cause loss of some sideband components or, alternatively, if the bandwidth is increased, some of the other sideband components may leak through. A case in which the latter technique is deliberately used to include a d.c. signal is known as vestigial sideband transmission (VSB) and is largely employed in television.

Since an SSBSC signal is composed of two DSBSC signals with their carriers and modulation 90° out of phase, it provides another way of generating an SSBSC signal and is called the phase-shift method. It is shown in Fig. 2.12 with its two balanced modulators, one of which has its carrier and modulation shifted by 90° relative to the other, and the combined outputs yield the SSBSC signal.

If v_1 and v_2 are the output voltages of the modulators, we have

$$v_1 = \frac{mV_c}{2}[\cos(\omega_c - \omega_m)t - \cos(\omega_c + \omega_m)t]$$

$$\begin{aligned} v_2 &= \frac{mV_c}{2}[\cos\{(\omega_c t + \pi/2) - (\omega_m t + \pi/2)\} \\ &\quad - \cos\{(\omega_c t + \pi/2) + (\omega_m t + \pi/2)\}] \\ &= \frac{mV_c}{2}[\cos(\omega_c - \omega_m)t + \cos(\omega_c + \omega_m)t] \end{aligned}$$

or

$$v_o = v_1 + v_2 = mV_c[\cos(\omega_c - \omega_m)t]$$

which is the lower sideband only.

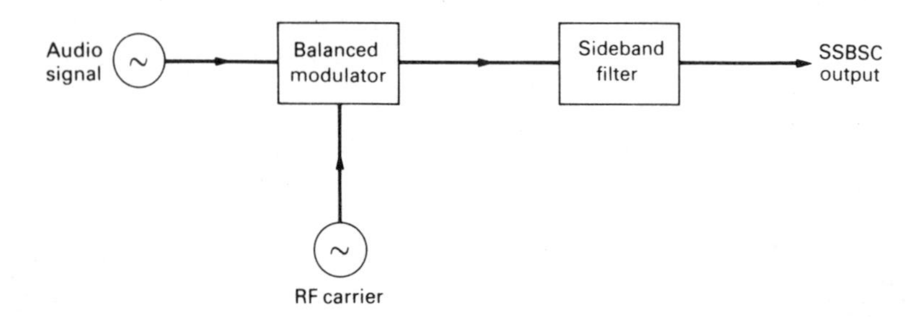

Fig. 2.11

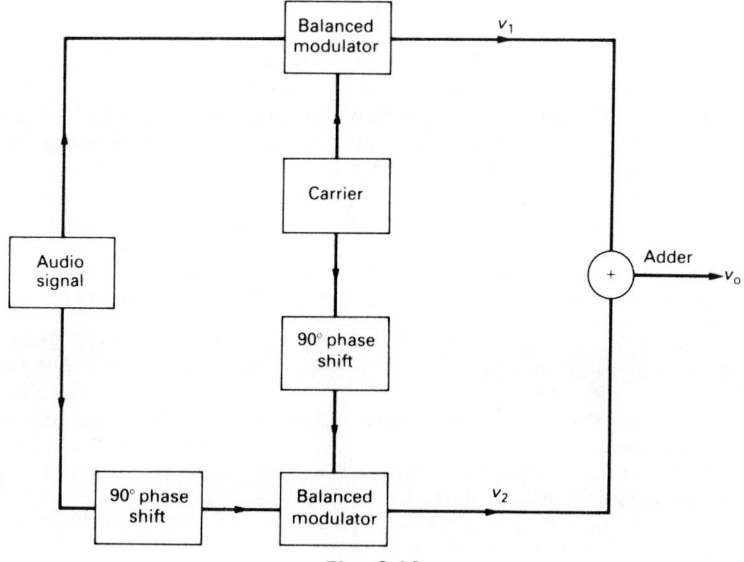

Fig. 2.12

A block diagram of a typical SSB transmitter is given in Fig. 2.13.

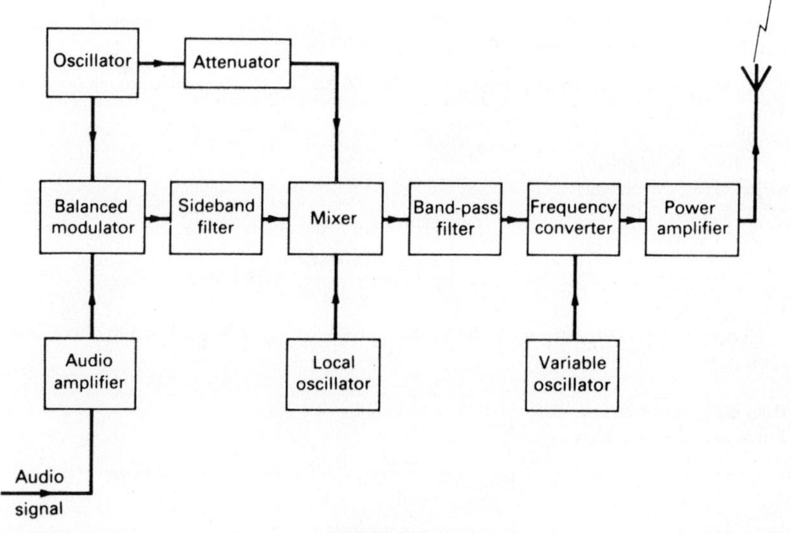

Fig. 2.13

Example 2.2

Discuss the reasons which lead to the use of double-sideband amplitude modulation for broadcast radio transmission, and single-sideband reduced carrier working for point-to-point transmission.

An AM transmitter has an output of 24 kW when modulated to a depth of 100%. Determine from first principles the power output (a) when the carrier is unmodulated and (b) when after modulation to a depth of 60% one sideband is suppressed and the carrier component is reduced by 26 dB. (U.L.)

Solution

The main reasons for using double-sideband AM for broadcast purposes are:

1. It simplifies equipment at the transmitter and receiver, thus reducing costs especially when a large number of receivers are involved.
2. An envelope detector can be used at the receiver. It is simple to design and does not call for any critical adjustment.

The main reasons for using single-sideband (reduced carrier) working for point-to-point communication are:

1. Only a restricted number of receivers are involved and so more expensive and sophisticated SSB equipment may be used.
2. There is a considerable saving of power which is especially economical for amateur enthusiasts who use SSB operation.
3. There is also a saving in bandwidth since only one sideband is transmitted. This greatly eases the present congested frequency spectrum especially in the amateur and marine bands.

Problem

Let P_c be the carrier power of the AM signal. Then

$$\frac{\text{total power}}{\text{carrier power}} = \frac{(1 + m^2/2)P_c}{P_c} = (1 + m^2/2)$$

(a) 100% modulation

Since $m = 1$, we have

$$(1 + \tfrac{1}{2})P_c = 24 \times 10^3$$

or

$$P_c = \frac{24 \times 10^3}{1{\cdot}5} = 16\,\text{kW}$$

Hence, the power output is the carrier power of 16 kW only when there is no modulation.

(b) 60% modulation

Since $m = 0{\cdot}6$, we have

$$\text{power in one sideband} = \frac{m^2 P_c}{4} = \frac{0{\cdot}36}{4} \times 16 \times 10^3 = 1440\,\text{W}$$

If P_{rc} is the power in the *reduced* carrier, we have

$$10 \log_{10} P_c/P_{rc} = 26$$

or
$$P_c/P_{rc} = 398$$

and
$$P_{rc} = \frac{16 \times 10^3}{398} = 40\ \text{W}$$

Hence
$$\text{total power output} = 1440\ \text{W} + 40\ \text{W} = 1480\ \text{W}$$

Example 2.3

Explain the single-sideband suppressed carrier system of amplitude modulation and discuss its merits. Why is a double-sideband suppressed carrier system not practicable?

The output stage of a certain radio transmitter can deliver a maximum of 10 kW (mean) radio-frequency power into the aerial. The transmitter is modulated to a depth of 40% by a sinusoidal signal. Compare the possible power in the sidebands when the carrier and both sidebands are radiated with that when the single-sideband suppressed carrier system is employed. (U.L.)

Solution

The answer to the first part of the question will be found in Section 2.5(b).

Merits

1. Considerable saving of carrier power and one sideband power.
2. Economy in bandwidth which reduces frequency congestion.
3. Distortion due to carrier fading is avoided.

DSBSC system

This system is not very practicable because the modulation in the received signal requires the re-insertion of the carrier at the receiving end. The carrier must be of the correct frequency and phase as the original carrier, in order that the received signal may be correctly demodulated. This calls for considerable sophistication at the receiver, which must be designed to provide the proper carrier, and this increases receiver costs. Moreover, if the phase is incorrect, the modulation is lost when the phase error approaches 90°.

Problem

Let P_c be the carrier power and m be the modulation factor. The power in the modulated carrier is given by $(1 + m^2/2)P_c$. Hence

$$(1 + m^2/2)P_c = 10\,000$$

with
$$\left[1 + \frac{(0{\cdot}4)^2}{2}\right]P_c = 10\,000$$

or
$$P_c = \frac{10\,000}{1{\cdot}08} = 9250\ \text{W}$$

AM system

$$\text{power in the sidebands} = (m^2/2)P_c = \frac{(0{\cdot}4)^2}{2} \times 9250$$

or
$$\text{sideband power} = 740\ \text{W}$$

SSBSC system

The maximum amplitude of the SSBSC signal from Section 2.5(b) is mV_c and its *mean* power is given by

$$m^2P_c = (0{\cdot}4)^2 \times 10^4 = 1600\ \text{W}$$

Hence
$$\frac{\text{SSBSC power}}{\text{AM sideband power}} = \frac{1600}{740} = 2{\cdot}16$$

and so the SSBSC power is about 3 dB greater than the AM sideband power.

Example 2.4

Explain why single-sideband transmission is commonly used for multichannel line telephony and point-to-point radio circuits. Outline two methods of producing a single-sideband signal.

A double-sideband amplitude-modulated radio transmitter gives a power output of 5 kW when the carrier is modulated with a sinusoidal tone to a depth of 95 %. If, after modulation by a speech signal which produces an average modulation depth of 20 %, the carrier and one sideband are suppressed, determine the mean output power in the remaining sideband. (U.L.)

Solution

Single-sideband transmission is used in multichannel telephony for economy of power and bandwidth. There is a saving in carrier power and one sideband power. Moreover, adjacent channels can be used for other messages leading to economy of bandwidth. In the case of point-to-point transmission, this is usually at low-power level, e.g. microwave links, military communications, amateur radio, etc., and so the total transmitted power can be concentrated in one sideband only to give the signal-to-noise required at the receiver. In amateur radio, especially, the saving in transmitter power is extremely useful.

The two methods for generating SSB signals are given hereafter.

First method

The first method generates an SSB suppressed carrier signal by using a balanced modulator, followed by an appropriate filter to remove one of the sidebands. This is shown in Fig. 2.14. Successive modulation and filtering can be used to bring the sideband into the correct part of the spectrum as required for aligning adjacent channels in a multichannel system.

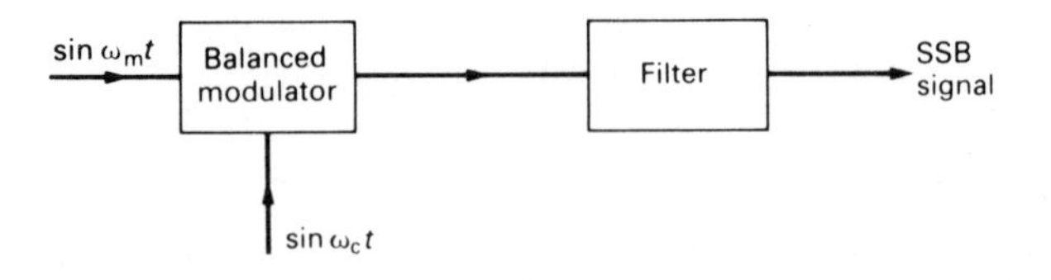

Fig. 2.14

Second method
The method employs the technique of phase discrimination. The carrier and modulation are shifted by 90° for one balanced modulator and are unchanged for another balanced modulator. The two outputs when combined yield a single-sideband signal. This is shown in Fig. 2.15.

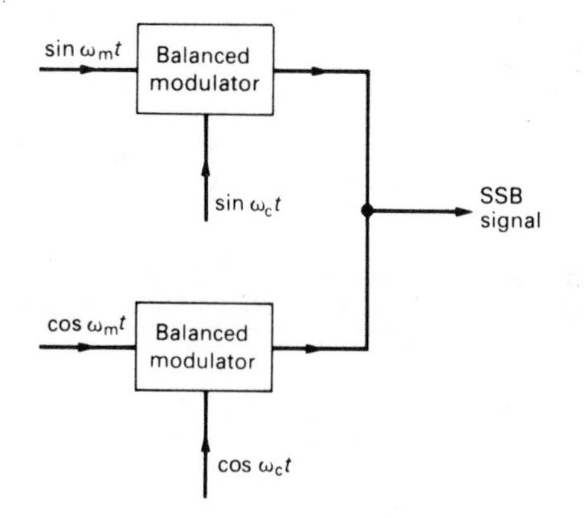

Fig. 2.15

Problem
Let P_c be the carrier power and P_o be the output power from the transmitter. Hence

$$P_o = P_c(1 + m^2/2) \quad \text{(for an AM signal)}$$

or
$$P_c = \frac{5000}{1 + \frac{(0{\cdot}95)^2}{2}} = \frac{5000}{1{\cdot}451} = 3450 \text{ W}$$

and
$$\text{mean SSB output power} = \frac{m^2 P_c}{4} = \frac{(0{\cdot}2)^2}{4} \times 3450 = 34{\cdot}5 \text{ W}$$

Comment
It is shown in Appendix A that the SSBSC signal can be expressed in terms of the *Hilbert transform* and is closely related to the phase discrimination method of generating an SSBSC signal.

(c) VSB system[9]
The video signal bandwidth for a 625-line television system is 6 MHz resulting in a modulation bandwidth of 12 MHz when AM is used. To economise bandwidth, a form of DSB is used in which a *vestige* of one sideband is

transmitted along with the whole of the other sideband. This reduces the overall bandwidth to about 8 MHz.

The transmitter characteristic is depicted in Fig. 2.16 where the vestige of the lower sideband includes transmission of the d.c. signal, since it represents the average brightness of the picture and is important picture detail. Consequently, on transmission, the lower frequency components of the vestigial sideband signal are overemphasised and so the receiver is given a response characteristic which reduces the low-frequency components, and this restores the information content on reception to its original balance prior to transmission. This is shown in Fig. 2.16.

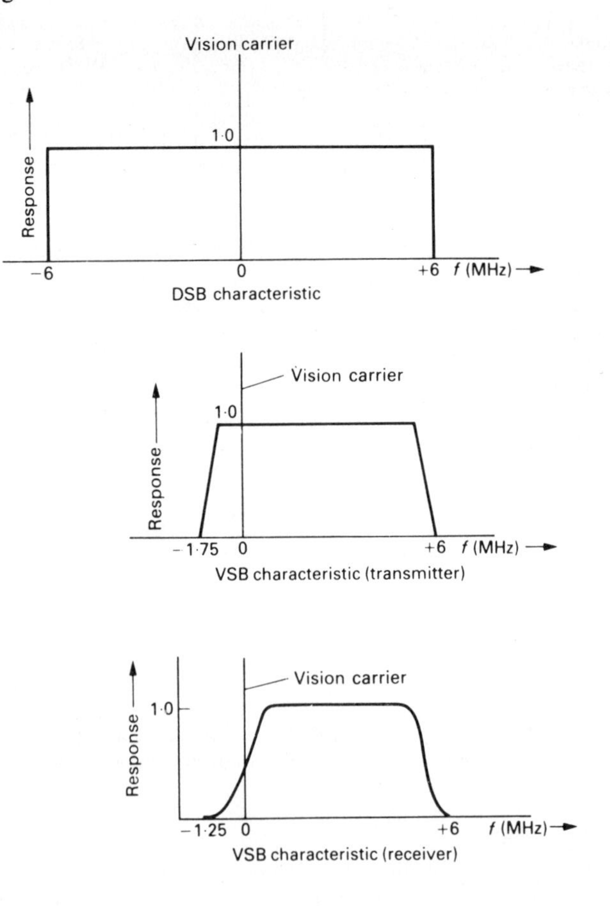

Fig. 2.16

Example 2.5

In what communication system would (a) single-sideband transmission or (b) vestigial sideband transmission be preferred to double-sideband transmission?

Assuming the signal to be transmitted may be represented by a single sine wave, find an approximate expression for the distortion present in the detected output of a vestigial sideband system in terms of the depth of modulation and the ratio of the magnitudes of the two sidebands. What practical significance has this calculation? (C.E.I.)

Solution

Single-sideband transmission is preferred in systems requiring a minimal bandwidth, as in multichannel carrier telephony, or point-to-point communications, as in amateur radio, where frequency bands are congested and power economy is important.

Vestigial sideband transmission is preferred in systems requiring a wide bandwidth, such as in television. It retains the advantages of DSB transmission, yet provides useful bandwidth economy.

VSB demodulation

The answer to this part of the question is given in Section 6.1.

(d) ISB system[10]

In certain applications, both sidebands of a DSB system are used, but each sideband carries a *different* message. The signal may be transmitted with or without a pilot carrier and is called *independent sideband* transmission. The process consists in producing two separate SSB signals. One message is superimposed on one of the sidebands and another message on the other sideband. The sidebands are arranged to occupy a 6 kHz bandwidth on either side of the carrier frequency. The two independent sidebands are then mixed with or without a carrier in a hybrid circuit. The arrangement is shown in Fig. 2.17.

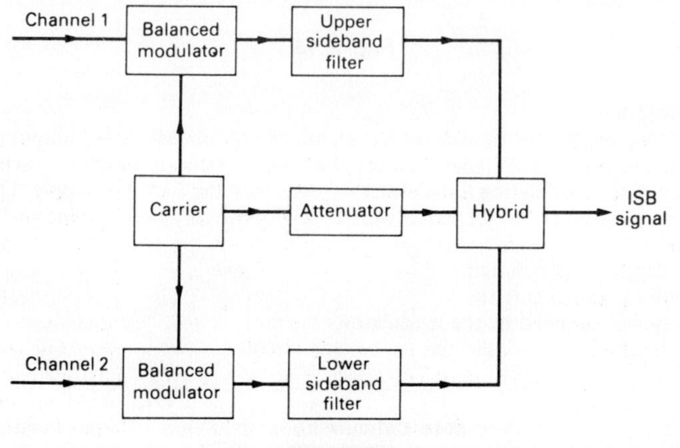

Fig. 2.17

2.6 AM transmitter[11]

A typical schematic arrangement of an AM transmitter is shown in Fig. 2.18. The audio signal from a microphone is amplified by a voltage amplifier and then by an audio-power amplifier for driving the modulator stage. The RF carrier wave is obtained from a stable crystal oscillator and for high stability it is separated by means of a buffer amplifier. The circuit uses low-level modulation and so the modulated carrier is amplified in a high-power linear RF amplifier prior to being fed to the aerial for transmission. A suitable d.c. power supply is also required to provide the energy for the transmitter. A typical medium-wave transmitter operates at a power level of about 50 kW for broadcasting speech and music.

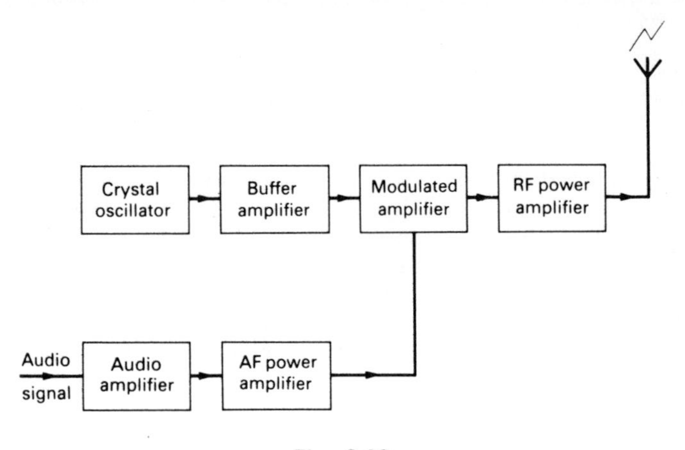

Fig. 2.18

Example 2.6

An amplitude-modulated transmitter has an anode-modulated class-C output stage in which an audio-frequency sine wave of 3 kV peak value is developed across the secondary of the modulating transformer in series with the 5 kV HT supply. The stage has an anode efficiency of 75 % and delivers 1·5 kW of carrier power into the tank circuit. Calculate

(a) the depth of modulation,
(b) the mean anode current,
(c) the power supplied by the modulator,
(d) the total RF power delivered to the tank circuit.

Solution

(a) Depth of modulation $= \dfrac{\text{peak audio swing}}{\text{peak carrier voltage}}$

or $$m = \frac{3000}{5000} = 0{\cdot}6$$

(b) $$\text{d.c. power} = \frac{1500}{0{\cdot}75} = 2000\ \text{W}$$

Also $$\text{d.c. power} = I_{\text{mean}} \times \text{d.c. voltage}$$

or $$I_{\text{mean}} = \frac{2000}{5000} = 0{\cdot}4\ \text{A}$$

(c) $$\text{Power supplied by modulator} = \frac{\text{sideband power}}{\text{efficiency}}$$

Also $$\text{sideband power} = m^2/2 \times \text{carrier power}$$

or $$\text{sideband power} = 0{\cdot}36/2 \times 1500 = 0{\cdot}27\ \text{kW}$$

so $$\text{modulator power} = 0{\cdot}27/0{\cdot}75 = 0{\cdot}36\ \text{kW}$$

(d) $$\text{Total power} = \text{carrier power} + \text{sideband power}$$
$$= 1{\cdot}5\ \text{kW} + 0{\cdot}27\ \text{kW}$$

or $$\text{RF power supplied} = 1{\cdot}77\ \text{kW}$$

Assumptions
1. The RF output varies in proportion to the anode voltage.
2. There are no losses in the modulating transformer.

3

Frequency modulation

The process of varying the frequency of a carrier wave in proportion to a modulating signal is known as frequency modulation (FM).[12] The carrier amplitude of an FM wave is kept constant during modulation and so the power associated with an FM wave is constant. During modulation, the carrier frequency increases when the modulating voltage increases positively and it decreases when the modulating voltage becomes negative. This is illustrated in Fig. 3.1.

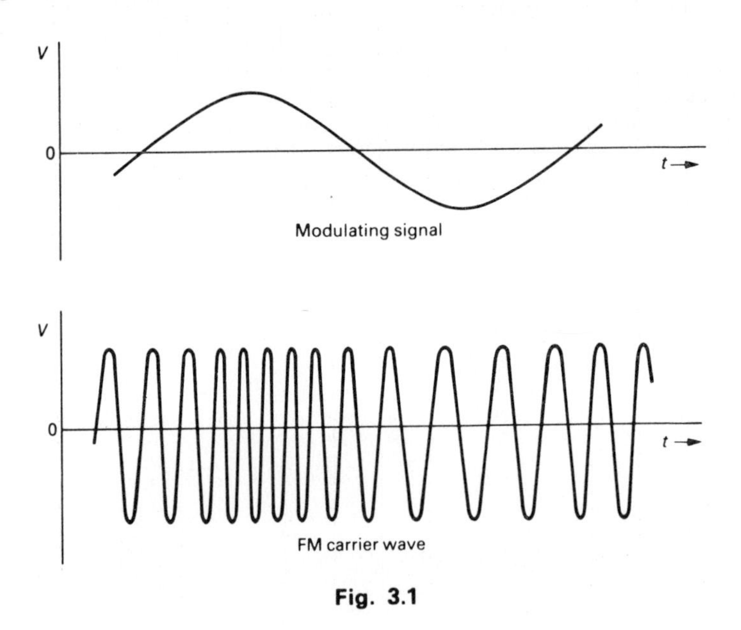

Fig. 3.1

To obtain an expression for an FM wave, let the instantaneous carrier wave be represented by

$$v_c = V_c \sin \omega_i t = V_c \sin 2\pi f_i t$$

where f_i is the instantaneous frequency. For a positive increase in frequency, we have

$$f_i = f_c + \Delta f_c \sin \omega_m t$$

where f_c is the carrier frequency and Δf_c is the *frequency deviation* of the carrier wave, due to the modulating signal of frequency f_m.

If the instantaneous carrier phase is ϕ_i, then

$$\frac{1}{2\pi}\frac{d\phi_i}{dt} = f_i = f_c + \Delta f_c \sin \omega_m t$$

or

$$\frac{d\phi_i}{dt} = 2\pi f_i = \omega_c + 2\pi \Delta f_c \sin \omega_m t$$

By integration and a correct choice of the phase angle, we obtain

$$\phi_i = \omega_c t - \frac{\Delta f_c}{f_m} \cos \omega_m t$$

or

$$\phi_i = \omega_c t - m_f \cos \omega_m t$$

where $m_f = \Delta f_c / f_m$ is called the modulation index.

Since $v_c = V_c \sin \phi_i$, we obtain

$$v_c = V_c \sin [\omega_c t - m_f \cos \omega_m t]$$

which represents an FM carrier wave.

3.1 The FM spectrum

Expanding v_c yields

$$v_c = V_c [\sin \omega_c t \cos (m_f \cos \omega_m t) - \cos \omega_c t \sin (m_f \cos \omega_m t)]$$

Now

$$\cos (m_f \cos \omega_m t) = J_0(m_f) - 2J_2(m_f) \cos 2\omega_m t + 2J_4(m_f) \cos 4\omega_m t - \ldots$$

$$\sin (m_f \cos \omega_m t) = 2J_1(m_f) \cos \omega_m t - 2J_3(m_f) \cos 3\omega_m t + \ldots$$

The coefficients $J_n(m_f)$ are Bessel functions of the first kind and order n. They are generally tabulated and a typical plot is shown in Fig. 3.2.

Substituting into v_c yields the result*

$$v_c = V_c [J_0(m_f) \sin \omega_c t - J_1(m_f) \{\cos (\omega_c + \omega_m)t + \cos (\omega_c - \omega_m)t\} - J_2(m_f) \{\sin (\omega_c + 2\omega_m)t + \sin (\omega_c - 2\omega_m)t\} + \ldots]$$

which reveals an infinite set of sidebands whose amplitudes are determined by

* See Appendix B.

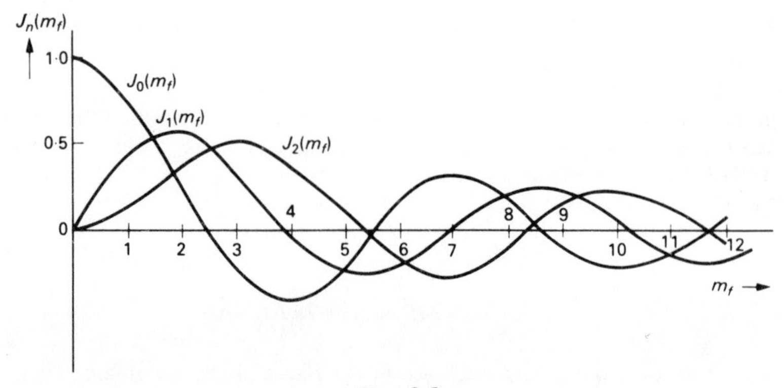

Fig. 3.2

the Bessel functions $J_0(m_f)$, $J_1(m_f)$, etc. Typical plots for $m_f = 0{\cdot}2$ and $m_f = 5{\cdot}0$ are shown in Fig. 3.3.

Fig. 3.3 shows that, when m_f is small, there are few sideband frequencies of large amplitude and, when m_f is large, there are many sideband frequencies but with smaller amplitudes. Hence, in practice, it is only necessary to consider a finite number of *significant* sideband components whose amplitudes are greater than about 4% of the unmodulated carrier. In practical FM systems, the frequency deviation is largely determined by the available bandwidth. Most FM broadcast stations use a frequency deviation of ± 75 kHz at the highest modulating frequency $f_h = 15$ kHz, which yields a *deviation ratio** of

$$\delta = \Delta f_c / f_h = \frac{75 \times 10^3}{15 \times 10^3} = 5$$

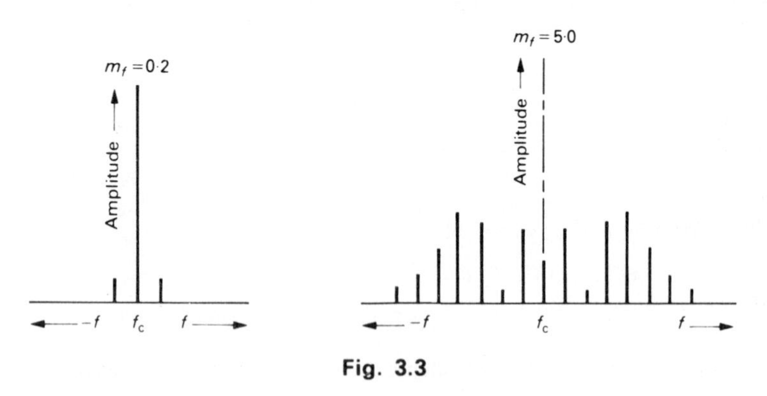

Fig. 3.3

* The deviation ratio is the value of the modulation index at the highest modulating frequency.

3.2 Phasor representation

A carrier wave of constant frequency f_c can be represented as a rotating phasor OA with constant angular velocity ω_c as shown in Fig. 3.4. If its frequency is slightly increased or decreased, the phasor would rotate slightly faster or slower than ω_c. Hence, relative to ω_c, the phasor OA would advance to position OB or be retarded to position OC. This amounts to varying the angle θ and so FM is a form of *angle* modulation and the phasor OA traces out the arc BAC.

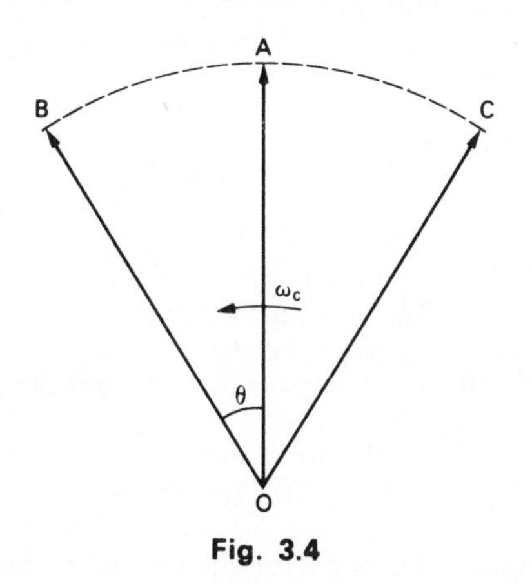

Fig. 3.4

Example 3.1

In the absence of an input signal, the carrier output from a distortionless frequency modulator has a frequency of 12 MHz and an amplitude of 5·0 V peak. An input signal causes a frequency deviation of 25 kHz per volt. Derive an expression for the modulated wave at the output when the signal $v = 1{\cdot}5 \sin 6280t$ volts is applied at the input. Then deduce

(a) the peak phase deviation of the modulated carrier,
(b) the number of times in each second that this deviation occurs,
(c) the peak frequency and phase deviations if the signal frequency is halved,
(d) the peak frequency deviation if the signal amplitude is doubled. (U.L.)

Solution

The expression for an FM wave was derived earlier as

$$v_c = V_c \sin\left[\omega_c t - m_f \cos \omega_m t\right]$$

where $m_f = \Delta f_c / f_m$ is the modulation index. Here

$$V_c = 5{\cdot}0 \text{ volts}$$
$$\omega_c = 2\pi f_c = 2\pi \times 12 \times 10^6 \text{ rads/s}$$
$$\omega_m = 6280 \text{ rads/s}$$

or

$$v_c = 5 \sin[24\pi\, 10^6 t - m_f \cos 6280t]$$

Problem

(a) Equating m_f to the peak phase deviation $d\phi$ we obtain

$$d\phi = \Delta f_c / f_m$$

Here $\Delta f_c = 25 \times 10^3 \times 1{\cdot}5\ \text{Hz}$

$f_m = 6280/2\pi = 10^3\ \text{Hz}$

Hence

$$d\phi = \frac{25 \times 10^3 \times 1{\cdot}5}{10^3} = 37{\cdot}5\ \text{rads}$$

(b) The peak deviation $d\phi$ occurs *twice* in each modulating period. Hence, its rate of occurrence is 2×10^3 times per second.

(c) The peak frequency deviation is unaltered and remains as 37·5 kHz. The phase deviation is doubled and becomes 75 rads.

(d) The peak frequency deviation is doubled and becomes $2 \times 37{\cdot}5$ kHz or 75 kHz.

3.3 Narrowband FM

From a plot of the Bessel functions shown earlier in Fig. 3.2, it will be observed that, for small values of m_f ($m_f < 1$), there is only one pair of significant sidebands when $m_f = 0{\cdot}1$ and two pairs of significant sidebands for $m_f = 0{\cdot}5$. These cases correspond to narrowband FM.

In the case of $m_f = 0{\cdot}1$, the carrier and first pair of sidebands yield the resultant phasor OD in Fig. 3.5(a). This result is similar to AM except that the lower sideband phasor AB is reversed due to its negative sign. Hence, we obtain essentially phase modulation of phasor OA through a small angle θ, together with some amplitude modulation at twice the audio frequency, i.e. $2f_m$. This signal can therefore be detected by an AM receiver.

To cancel out the residual amplitude modulation, it requires the addition of a second pair of sidebands which are in phase with the carrier and rotating at a frequency $2f_m$ relative to the carrier. This is shown in Fig. 3.5(b), where the resultant phasor OD is of constant amplitude and yields phase modulation through an angle ϕ. This phase modulation accounts for the frequency modulation present.

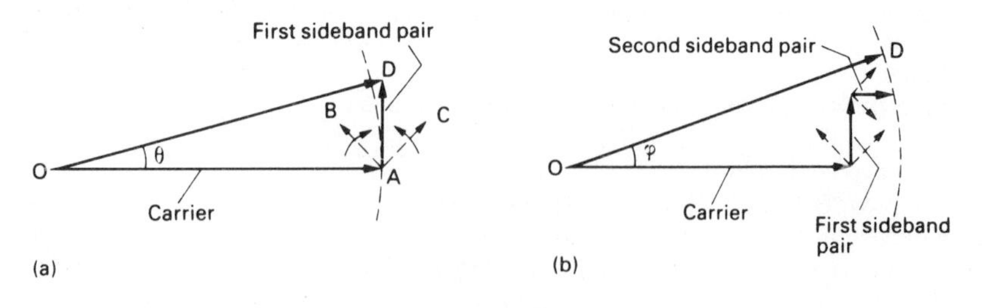

Fig. 3.5

3.4 Broadband FM

For values of $m_f \gg 1$, the sidebands cover a wide spectrum but their amplitudes decrease. Hence, the number of significant sidebands does not depend very much on m_f when the number of sidebands is greater than about 10. In this case, the practical bandwidth may be taken as approximately equal to $2(\Delta f_c + f_h)$, where Δf_c is the frequency deviation of the system and f_h is the *highest* modulating frequency. For the case of $m_f = 5$, $\Delta f_c = 75$ kHz, and $f_h = 15$ kHz, we obtain a bandwidth of $2(75+15)$ kHz $= 180$ kHz for a typical FM broadcast system.

This is considerably larger than the bandwidth used with the corresponding AM system and accounts for the inherent wideband nature of FM. However, the use of a large bandwidth leads to a considerable improvement in signal-to-noise ratio (see Section 3.7) which is one of the main advantages of FM compared to AM. Narrowband FM does not possess this property due to its small frequency deviation Δf_c. It can be shown that, the larger the value of Δf_c, the greater is the signal-to-noise ratio improvement.

Example 3.2

Develop the spectrum of a carrier signal modulated sinusoidally in frequency. Show that for small deviation ratios all but the carrier and one pair of sidebands can be neglected, and sketch the resultant vector diagram. For larger deviation ratios, show that the inclusion of a second pair of sidebands can result in a more accurate vector diagram.
(C.E.I.)

Solution

The answer to the first part of the question is given at the beginning of this chapter in Section 3.1.

The answers to the remaining parts of the question are given in Section 3.3 and Section 3.4 respectively and the phasor diagrams in Fig. 3.5.

3.5 FM generation

A *direct* method for producing an FM wave is to vary the frequency of the carrier by means of the modulating signal. This can be done by using a non-linear device, such as a valve or transistor, to vary the capacitance across the tuned circuit of an oscillator. Alternatively, an *indirect* method using phase modulation will be considered in the next chapter.

Varactor modulator

A typical circuit uses the variable capacitance of a p–n junction semiconductor which is known as a varactor diode.

When a d.c. voltage is applied in the reverse direction across the p–n junction shown in Fig. 3.6, the charges are drawn away from the junction leaving a depletion layer at the junction because it is depleted of charge carriers. This gives rise to a capacitance which appears across the junction between the separated charges and it may be varied by varying the applied d.c. voltage. The capacitance at the junction varies approximately according to the expression

$$C_j \propto 1/\sqrt{V}$$

where V is the applied reverse voltage.

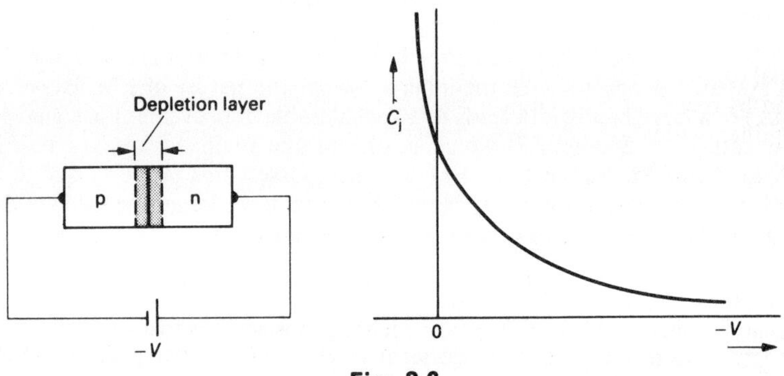

Fig. 3.6

Silicon diodes are usually made to have a capacitance of between 150 and 200 pF with 1 volt applied, which decreases to about 50 pF with 10 volts applied. The average variation is about 10 to 15 pF per volt of reverse bias.

In the typical varactor modulator shown in Fig. 3.7, the Zener diode stabilises the d.c. supply so that the mean oscillator frequency is not altered by

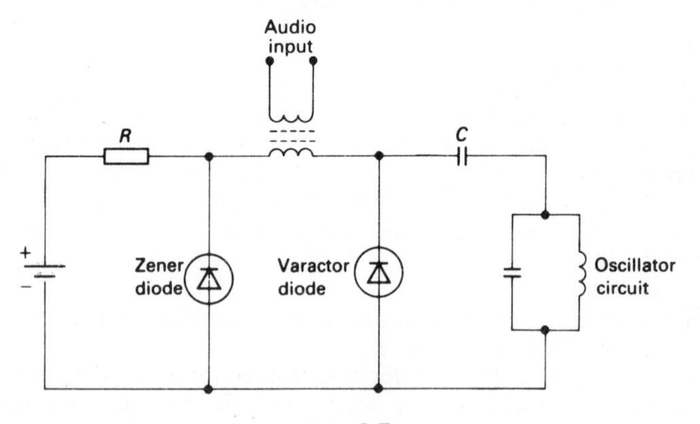

Fig. 3.7

supply voltage fluctuations. The varactor capacitance C_j is varied by the modulating voltage v_m and the oscillator uses collector–emitter feedback. The coupling capacitor provides d.c. isolation for the oscillator circuit.

3.6 FM transmitter

Fig. 3.8 is a block schematic diagram of an FM transmitter employing a varactor modulator. The audio signal is amplified in the AF stages and drives the varactor modulator. The latter varies the frequency of an *LC* oscillator, whose *centre* frequency is stabilised by crystal control via an AFC loop. The initial carrier frequency is multiplied several times to bring it up into the VHF band by means of frequency multipliers. The output drives class-C RF power amplifiers with high efficiency, to give an FM output of a few kilowatts.

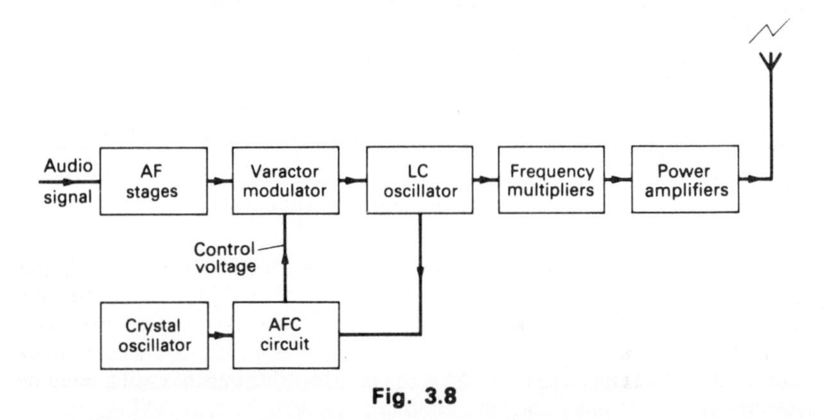

Fig. 3.8

Example 3.3

Use a diagram to explain the special characteristic of a varactor diode. Describe, with the aid of a simple circuit diagram, the operation of a varactor diode frequency modulator.

An oscillator operating at 100 MHz has a 75 pF capacitor in its tuning circuit. What total capacitance swing must the varactor supply to allow the modulator to have a 80 kHz peak deviation? (C.G.L.I.)

Solution

The answer to the first part of the question is given in Section 3.5.

Problem

$$f = \frac{1}{2\pi\sqrt{LC}}$$

where f is the resonant frequency, and L and C are the tuned circuit inductance and capacitance respectively.

If the increase in tuning capacitance is ΔC, let the frequency decrease by Δf such that

$$f-\Delta f=\frac{1}{2\pi\sqrt{L(C+\Delta C)}}$$

or

$$\frac{f}{f-\Delta f}=\sqrt{\frac{L(C+\Delta C)}{LC}}=(1+\Delta C/C)^{1/2}$$

and

$$\frac{f-\Delta f}{f}=(1+\Delta C/C)^{-1/2}$$

with

$$1-\Delta f/f=1-\frac{1}{2}\left(\frac{\Delta C}{C}\right)+\ldots$$

by the binomial theorem if $\Delta C \ll C$.

Hence

$$\Delta f/f\simeq\frac{1}{2}\left(\frac{\Delta C}{C}\right)$$

or

$$\Delta C\simeq 2\frac{\Delta f\times C}{f}$$

Substituting for $\Delta f = 8\times 10^4$ Hz, $f=100$ MHz, and $C=75$ pF yields

$$\Delta C=\frac{2\times 8\times 10^4\times 75\times 10^{-12}}{100\times 10^6}=0{\cdot}12\,\text{pF}$$

Since the total frequency swing is ± 80 kHz, the total capacitance swing is $\pm 0{\cdot}12$ pF.

3.7 Interference and noise[13]

One of the great advantages of FM is the considerable improvement in signal-to-noise ratio obtainable compared to that obtainable from an AM system. The effects of interference and noise are somewhat similar and can be examined in much the same way.

Interference

An interfering signal may be either very close to the carrier of an FM station, same channel interference, or it may be in an adjacent band, adjacent channel interference. The effect of such interference depends upon the amount of frequency modulation it produces in an FM receiver, as the receiver is sensitive only to frequency modulation.

In general, let the interfering signal frequency be f_i and the FM carrier frequency be f_c, where $f_i > f_c$. Fig. 3.9 shows how the interfering signal will beat with the FM carrier and it can be regarded as rotating *relative* to the carrier with the *difference* frequency $f_d = f_i - f_c$. It is assumed that the carrier amplitude is larger than that of the interfering signal, as this assumption is quite justified in practice.

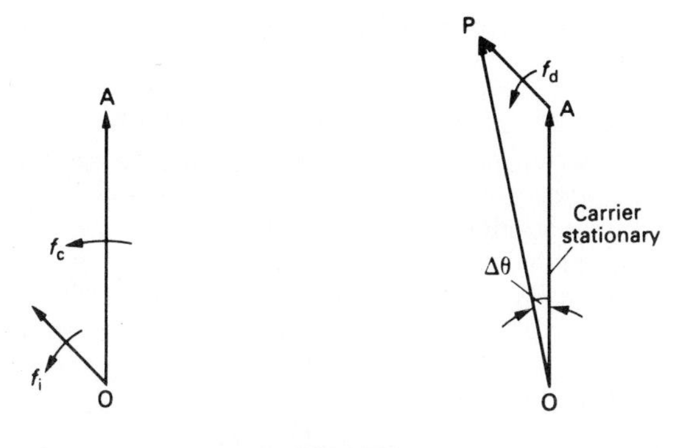

Fig. 3.9

The resultant phasor OP in Fig. 3.9 produces both AM and PM. The AM can be removed by a limiter and the only effect to be considered is that of PM. It can be shown that a frequency shift Δf due to a phase shift $\Delta\theta$ is given by $\Delta f = f_d \Delta\theta$, where f_d is the difference frequency of the interference. Since the frequency shift determines the output volts produced in an FM receiver, we observe that, when f_d is small (same channel interference), Δf is small and, when $f_d = 0$, $\Delta f = 0$.

Hence, the interfering effect is zero when it is on the same frequency as the FM station. In this case, the FM receiver[14] captures the stronger of the two signals which is known as the *capture effect*.[15] As f_d increases (adjacent channel interference), Δf increases, but the receiver response falls off due to the selectivity characteristics of its tuned circuits and the interference will have less effect on the receiver. Any interfering signal frequency greater than 15 kHz from the wanted carrier need not be considered as the audio response of the receiver is designed to fall off rapidly beyond 15 kHz.

Such adjacent channel interference which may be due to neighbouring FM stations is minimised by the use of a guard band on either side of the carrier's sidebands and by suitably spacing the carrier frequencies of nearby FM stations, such that their guard bands do not overlap. The interference is also reduced by the use of VHF carrier frequencies which can only be received by line-of-sight propagation and so the separation of FM stations, by distances of about 80 km, helps to reduce the effect.

Noise

Noise effects are due to both static and random noise. Static noise due to atmospherics tends to diminish somewhat at the VHF frequencies used for FM carriers, compared to the greater amount of static noise at the lower frequencies of the medium- and short-wave bands employed for AM reception. Furthermore, for static noise voltages smaller than that of the wanted carrier, the frequency modulation produced in an FM receiver is small, as was shown earlier for interference.

In the case of random noise, the noise voltages are random in phase and beat with the FM carrier. The beat or difference frequency $f_d = |f_n - f_c|$, where f_n is the noise frequency which may be greater or less than f_c. Hence, f_d can vary from 0 to any large value (positive or negative). However, only values up to ± 15 kHz relative to f_c produce audible noise in the FM receiver. The amount of noise power produced can be evaluated by noting that both AM and FM are produced as shown in Fig. 3.10. The AM is removed by the limiter and has no effect on the FM receiver, while the remaining FM gives rise to noise voltages in the FM receiver which are proportional to f_d, since $\Delta f = f_d \Delta\theta$ and $\Delta\theta$ is assumed constant for the analysis.

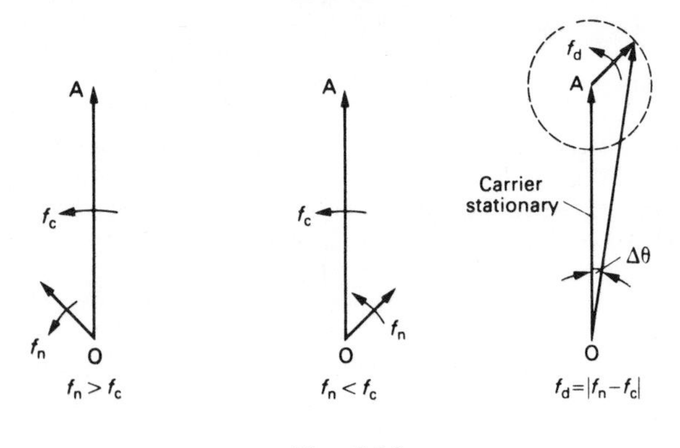

Fig. 3.10

In order to compare the effects of random noise on an AM and an FM receiver respectively, it is useful to assume that the AM and FM carriers are of equal power and modulated 100%, i.e. $m = 1$ for AM and $\Delta f = \pm 75$kHz for FM. Also, the audio bandwidths of both receivers are assumed to be 15 kHz for a fair comparison to be made. In the case of an AM envelope detector, the output voltage is independent of the noise frequency and is the same at each frequency. Using a normalised value of 1 volt for 100% modulation, we obtain the rectangular distribution shown in Fig. 3.11. For FM, the output noise voltage is *proportional* to f_d and we obtain the triangular distribution shown in Fig. 3.11. Due to symmetry, only positive values of f_d need to be considered.

To evaluate the relative noise powers for both cases, consider any noise component at frequency f_n, where $f_n > f_c$. The noise voltage it produces at $f_d = (f_n - f_c)$ is 1 volt for the case of AM and for FM it is y volts, where $y = f/f_2$. To obtain the total noise power in each case, it is necessary to consider first the

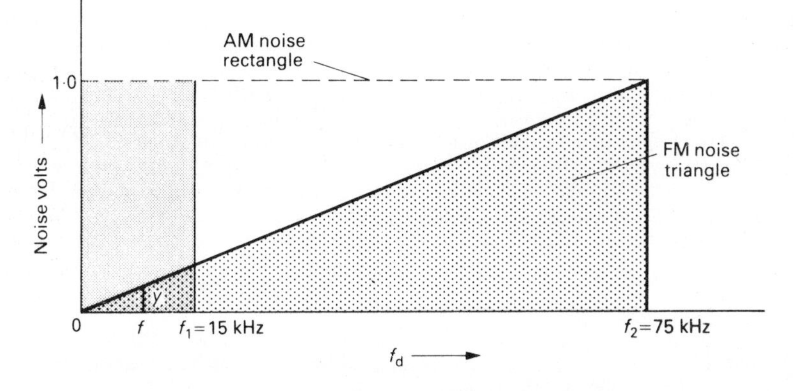

Fig. 3.11

noise power within a bandwidth df and to integrate it over the audible bandwidth from $f_d = 0$ to $f_d = f_1$, where $f_1 = 15$ kHz. Assuming a load resistance of 1 Ω, we obtain

$$\text{noise power from AM receiver} = \int_0^{f_1} 1^2 \, df = f_1 \quad \text{watts}$$

$$\text{noise power from FM receiver} = \int_0^{f_1} y^2 \, df = \int_0^{f_1} \frac{f^2 df}{f_2^2} = \frac{f_1^3}{3f_2^2} \quad \text{watts}$$

or

$$\frac{\text{AM noise power}}{\text{FM noise power}} = \frac{f_1}{f_1^3/3f_2^2} = 3\left(\frac{f_2}{f_1}\right)^2 = 3\delta^2$$

where $\delta = f_2/f_1$ is the deviation ratio of the FM system. In the case of FM broadcasting, $\delta = 5$ and we obtain

$$\frac{\text{AM noise power}}{\text{FM noise power}} = 3 \times 5^2 = 75$$

which amounts to an FM noise *improvement* of 75 times or 19 dB.

A further noise improvement is possible by the use of pre-emphasis and de-emphasis techniques. Since most FM noise received is at the higher audio frequencies, pre-emphasis circuits are used at the transmitter to boost the high frequencies in the modulating signal. On reception, the original signal balance is restored by de-emphasising the high frequencies to the same extent and this also de-emphasises a lot of high-frequency noise.

It is shown in Appendix C that another 4 or 5 dB improvement is possible and so the total noise improvement of FM over AM is about 23 dB. This amounts to a signal-to-noise improvement of 23 dB and is due to the greater bandwidth used by an FM system compared to that of an AM system.

Example 3.4
Compare and contrast amplitude and frequency modulation systems and discuss the differences at the transmitter and receiver.

The modulation in a frequency-modulated transmitter is achieved by variation of the tuning capacitance of an oscillator operating at a mean frequency of 3 MHz. The coil used in the parallel tuned circuit of the oscillator has an inductance of 10 μH. If the modulated waveform is frequency-multiplied to give an output of 120 MHz, with a maximum frequency deviation of 180 kHz, determine the change in value of capacitance to be produced by the modulating signal. (U.L.)

Solution

AM and FM comparison
1. In both systems, a carrier wave is modulated by an audio signal to produce a carrier and sidebands. The technique can be applied to various communication systems, such as telegraphy, telephony, etc.
2. Both systems use receivers based on the superheterodyne principle.
3. Special techniques applied to AM receivers, such as AGC, can also be applied to FM receivers.

AM and FM contrast
1. In AM, the carrier amplitude is varied whereas, in FM, the carrier frequency is varied.
2. AM produces two sets of sidebands and is a narrowband system. FM produces a large set of sidebands and is a broadband system.
3. FM gives a much better signal-to-noise ratio than does AM under similar operating conditions.
4. Though there are points of similarity in the transmitters and receivers, there are also noticeable differences.
5. FM systems are usually more sophisticated and expensive than AM systems.

Transmitters
Provision must be made for varying the carrier amplitude in an AM transmitter whilst for FM the carrier frequency is varied. AM and FM modulators are therefore essentially different in design. Moreover, FM can be produced by direct frequency modulation or indirectly by phase modulation. However, the FM carrier must be high and is usually in the VHF band, because it requires a large bandwidth which is not available in the lower congested bands.

Receivers
Though AM and FM receivers are basically the same, an FM receiver uses a limiter and discriminator (or a ratio detector) to remove AM variations and to convert frequency changes into amplitude variations respectively. Hence, FM receivers must have higher gain than AM receivers in order to operate the discriminator effectively. Finally, FM receivers give high-fidelity reproduction due to their larger audio bandwidth up to 15 kHz compared with about 8 kHz for AM receivers.

Problem

$$f_c = 3\,\text{MHz}$$

$$f_r = \frac{1}{2\pi\sqrt{LC}}$$

$$C = \frac{1}{9 \times 10^{12} \times 4\pi^2 \times 10 \times 10^{-6}} = 280\,\text{pF}$$

Initially $$\Delta f = \frac{180 \times 10^3}{120/3} = 4{\cdot}5\,\text{kHz}$$

Also $$f \pm \Delta f = \frac{1}{2\pi\sqrt{L(C \mp \Delta C)}}$$

and $$\delta C = 2C\frac{\delta f}{f} = 2 \times 280 \times 10^{-12} \times \frac{4{\cdot}5 \times 10^3}{3 \times 10^6}$$

or $$\delta C = 0{\cdot}84\,\text{pF}$$

Hence $$\text{total capacitance variation} = 2 \times 0{\cdot}84\,\text{pF} = 1{\cdot}68\,\text{pF}$$

3.8 FM stereo[11,16]

Stereophonic music can be broadcast over an FM station which uses a peak frequency deviation of ± 75 kHz. Since the audio bandwidth allocated to music is 15 kHz, the spectrum beyond 15 kHz can be used to accommodate another audio channel. The two audio channels are called the left and right channels (L and R signals) and they are combined in a matrix circuit to produce the (L + R) signal which is the ordinary mono signal and the (L − R) signal which carries the stereo information.

The (L + R) signal occupies the frequency band up to 15 kHz, while the (L − R) signal is transposed by a 38 kHz subcarrier to the frequency band between 23 kHz and 53 kHz. The subcarrier signal is obtained from a 19 kHz crystal oscillator which is doubled (squared) to 38 kHz and then modulated by the (L − R) signal in a balanced modulator to produce a DSBSC signal.

The (L + R) signal, the DSBSC signal, and a 19 kHz pilot carrier are added together to form a composite signal which is used to frequency-modulate the FM transmitter. Ninety per cent of the peak deviation ($\pm 67{\cdot}5$ kHz) is due to the audio information while the remaining ten per cent ($\pm 7{\cdot}5$ kHz) is due to the rest of the transmitted information. The spectral bands used and the encoder are shown in Fig. 3.12.

At the receiver, the demodulated signal and the pilot carrier, after separation and squaring, are fed into a synchronous detector to obtain the original (L + R) and (L − R) signals. At the output, the signals are added or subtracted by a matrix circuit to produce the L and R signals for driving separate loudspeakers. The decoder is shown in Fig. 3.13 and it should be observed that de-emphasis is applied *after* synchronous detection to avoid de-emphasising the high-frequency components of the stereo information in the received signal. In an alternative method, the gating principle is used.[11]

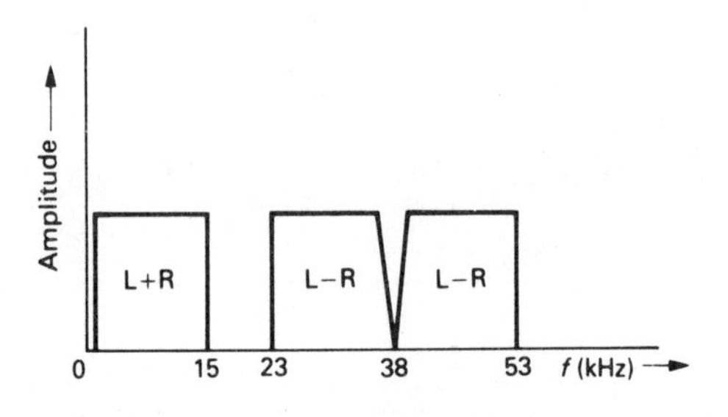

(a) Spectral bands

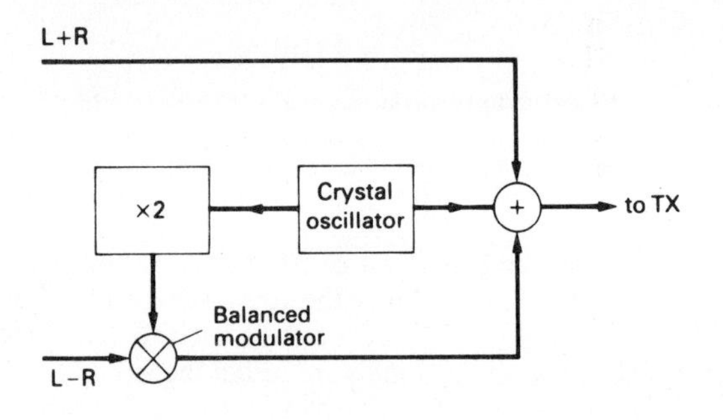

(b) Encoder

Fig. 3.12

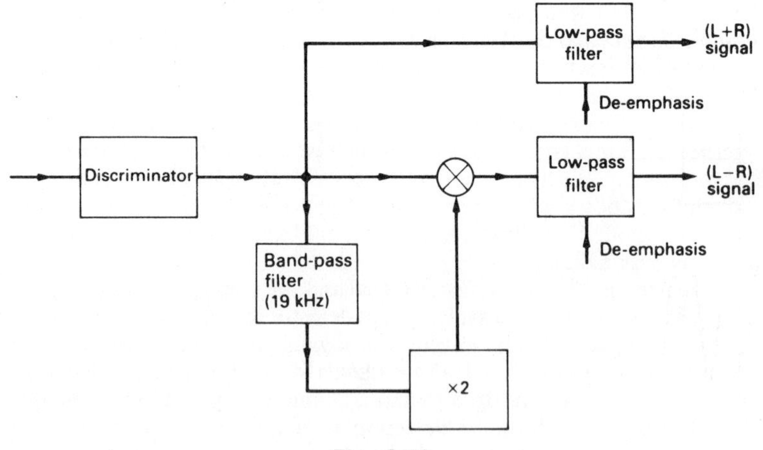

Fig. 3.13

4

Phase modulation

Angle modulation of a carrier also leads to another form of modulation known as phase modulation. Hence, both phase modulation and frequency modulation are produced during angle modulation and, as a consequence, FM and PM are very closely linked to one another. However, there are some important differences between PM and FM that warrant a separate study of PM.

In phase modulation, the *phase* of the carrier is varied proportionately by the modulating voltage, i.e. the phase shift increases with an increase in modulating voltage and vice versa. If v_i is the instantaneous unmodulated carrier voltage, then let

$$v_i = V_c \sin(\omega_c t + \phi)$$

where ϕ is any arbitrary phase of the carrier. If the modulating signal is of the form $v_m = V_m \sin \omega_m t$, where $\omega_m = 2\pi f_m$ and f_m is the audio frequency, let the carrier phase vary sinusoidally with a peak *phase deviation* $\Delta\phi$, then the phase-modulated carrier is given by

$$v_c = V_c \sin[\omega_c t + \phi + \Delta\phi \sin \omega_m t]$$

or

$$v_c = V_c \sin[\omega_c t + \Delta\phi \sin \omega_m t]$$

if initially ϕ is assumed to be zero.

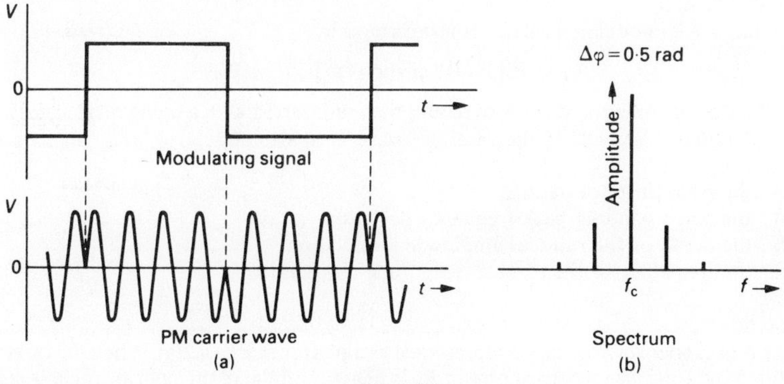

Fig. 4.1

The phase shifts in a PM carrier wave are easily shown for a square-wave modulating signal and are illustrated in Fig. 4.1(a). Here, the carrier-phase shifts by 180° whenever the modulating voltage changes abruptly in value. A typical spectrum for a *sinusoidal* modulating signal is shown in Fig. 4.1(b).

4.1 The PM spectrum

The expression for a PM carrier wave is very similar in form to that of an FM carrier wave if $\Delta\phi = \Delta f_c/f_m$. Hence, putting $\Delta\phi = m_p$, where m_p is called the modulation index, we obtain

$$v_c = V_c[J_0(m_p)\sin\omega_c t + J_1(m_p)\{\sin(\omega_c+\omega_m)t - \sin(\omega_c-\omega_m)t\} + J_2(m_p)\{\sin(\omega_c+2\omega_m)t + \sin(\omega_c-2\omega_m)t\} + \ldots]$$

The spectrum of a PM wave shown in Fig. 4.1(b) contains an infinite set of sidebands whose amplitudes are given by the Bessel functions $J_0(m_p)$, $J_1(m_p)$, etc. Consequently, PM and FM carrier waves would have identical spectra if $\Delta\phi = m_p = m_f$, where $m_f = \Delta f/f_m$. However, in PM, for practical reasons, $\Delta\phi$ is given a *fixed* maximum value and so, as f_m varies, the frequency deviation Δf of a PM wave also varies with the modulating frequency, so that $\Delta\phi = \Delta f/f_m$ remains constant.

This is quite different from FM in which Δf is constant and can be given a large value. As signal-to-noise ratio improvement depends upon a large Δf, this is the main reason why FM is preferred to PM in many practical applications. The essential difference therefore between PM and FM is the fact that, in the former case, Δf is proportional to f_m and must be restricted to keep $\Delta\phi$ small while, in the latter case, Δf is independent of f_m and can be quite large to give a good signal-to-noise ratio on reception.

Example 4.1
An amplitude-modulated carrier is represented by

$$e = E_c[1 + 0{\cdot}2\cos(\pi 10^3 t)]\sin(2\pi 10^7 t)$$

The carrier component is removed and then re-inserted with a phase displacement of $+\pi/2$ radians. Show that the resulting wave is phase-modulated, and calculate the following
(a) the peak phase deviation,
(b) the corresponding peak frequency deviation,
(c) the depth of the residual amplitude modulation,
(d) the frequency of the residual amplitude modulation. (U.L.)

Solution
The AM carrier wave is shown represented as a phasor in Fig. 4.2(a). When the carrier is shifted by $+\pi/2$, the resultant phasor E_R is phase-modulated through an angle $\pm\Delta\phi$ as shown in Fig. 4.2(b). Assuming $\Delta\phi$ is a small angle, we obtain

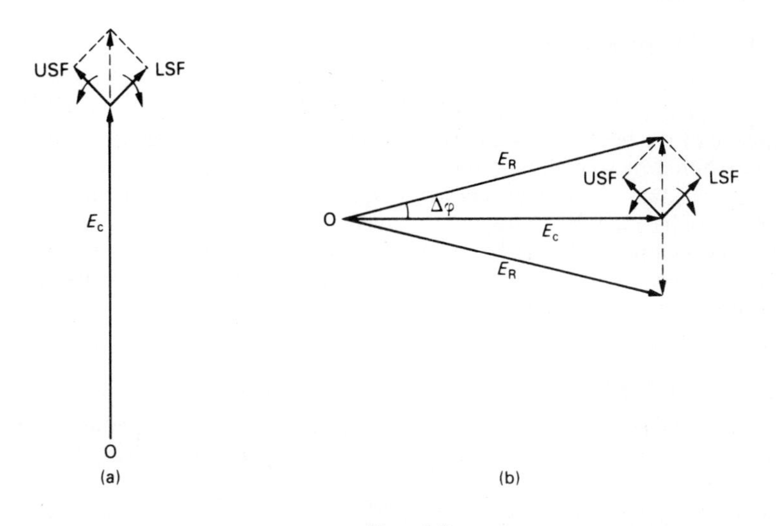

Fig. 4.2

(a) Peak phase deviation $= \Delta\phi$

$$\tan \Delta\phi = \frac{0{\cdot}2E_c}{E_c} = 0{\cdot}2$$

or
$$\Delta\phi \simeq 0{\cdot}2$$

since $\tan \Delta\phi \simeq \Delta\phi$ for small angles.

(b) The peak frequency shift Δf_c produced in a carrier wave of frequency f_c, by a phase deviation $\Delta\phi$, is given by

$$\Delta f_c = f_m \Delta\phi$$

or
$$\Delta f_c = (\pi\, 10^3/2\pi)(0{\cdot}2) = 100\,\text{Hz}$$

(c) The depth of the residual AM is given by

$$m = \frac{E_R - E_c}{E_c} = \frac{E_R}{E_c} - 1$$

Now
$$E_R = \sqrt{E_c^2 + (0{\cdot}2E_c)^2} = E_c\sqrt{1 + 0{\cdot}04} = E_c(1{\cdot}02)$$

Hence
$$m = \frac{E_R}{E_c} - 1 = 1{\cdot}02 - 1 = 0{\cdot}02$$

(d) The frequency of the residual AM is twice that of the modulation, since the carrier reaches a peak value *twice* in one cycle of modulation. Hence

$$\text{residual AM frequency} = 5 \times 500\,\text{Hz} = 1\,\text{kHz}$$

4.2 PM/FM generation

One of the main applications of phase modulation is in the indirect generation of an FM carrier wave, which is known as the Armstrong[17] system. The main advantage of the technique is the use of a very stable crystal oscillator whose phase is shifted by a small amount.

To ensure that the corresponding frequency shift is constant and independent of the modulating frequency (as for FM), the modulating signal must be modified before application to the phase modulator. This is done by using an integrating network as shown in Fig. 4.3, the output of which is then applied to the phase modulator.

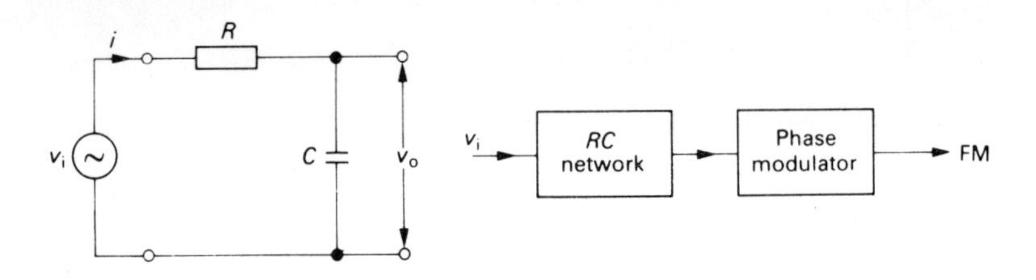

Fig. 4.3

For the integrating network shown in Fig. 4.3 we have

$$i = \frac{v_i}{R - \dfrac{j}{\omega_m C}}$$

$$v_o = i(-j/\omega_m C) = \frac{v_i}{R - \dfrac{j}{\omega_m C}} \left[-\frac{j}{\omega_m C} \right]$$

or

$$\frac{v_o}{v_i} = \frac{-j}{\omega_m RC - j}$$

with

$$\frac{|v_o|}{|v_i|} = \frac{1}{\sqrt{1 + \omega_m^2 R^2 C^2}}$$

If $\omega_m RC \gg 1$, then

$$\frac{|v_o|}{|v_i|} = \frac{1}{\omega_m RC}$$

with

$$|v_o| = k/f_m$$

where k is a constant of proportionality.

Since $\Delta f = f_m \Delta\phi$ and $\Delta\phi$ is proportional to $|v_o|$ in PM, hence

$$\Delta f = f_m k/f_m = k$$

which is a constant as in frequency modulation.

The basic feature of the Armstrong modulator is to phase-modulate a carrier which has been shifted 90° relative to the modulating component. The phase shift is kept small (< 0·5 rad) to reduce distortion and the consequent frequency shift Δf, being too small initially, is multiplied up to ± 75 kHz by frequency-multiplier stages. The phasor diagram is shown in Fig. 4.4 and the basic circuit in Fig. 4.5.

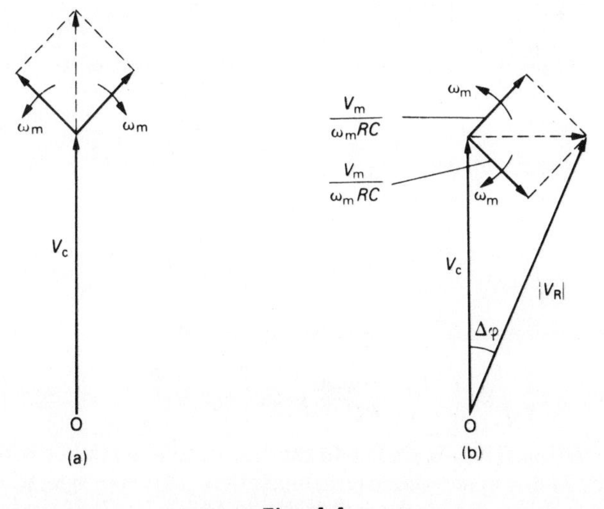

Fig. 4.4

To show that the output of the modulator is an FM wave, let the modulating signal be v_i, where $v_i = V_m \sin \omega_m t$, and the output from the integrating circuit be v_o, where

$$v_o = \frac{-jv_i}{\omega_m RC - j} = \frac{v_i}{1 + j\omega_m RC} = \frac{V_m \sin \omega_m t}{\omega_m RC \angle 90^\circ}$$

if $\omega_m RC \gg 1$. Hence

$$v_o = \frac{V_m}{\omega_m RC} \sin \omega_m t \angle -90^\circ = \frac{-V_m}{\omega_m RC} \cos \omega_m t$$

If this signal modulates a carrier $v_c = V_c \sin \omega_c t$, the sideband output of the balanced modulator is $v_s = kv_o \times v_c$

or
$$v_s = \frac{-kV_m V_c}{\omega_m RC} \sin \omega_c t \cos \omega_m t$$

where k is a constant of proportionality.

The addition of these components 90° out of phase with another carrier $v_c = V_c \sin \omega_c t$ gives the resultant phasor v_R shown in Fig. 4.4, where

$$v_R = v_c + jv_s = V_c \sin \omega_c t - \frac{kV_c V_m}{\omega_m RC} \cos \omega_c t \cos \omega_m t$$

since $j \sin \omega_c t = \cos \omega_c t$.

Substituting

$$\frac{kV_c V_m}{\omega_m RC} \cos \omega_m t = A \sin \Delta\phi \quad \text{and} \quad V_c = A \cos \Delta\phi$$

yields

$$v_R = A \sin \omega_c t \cos \Delta\phi - A \cos \omega_c t \sin \Delta\phi = A \sin (\omega_c t - \Delta\phi)$$

where

$$A = V_c \sqrt{1 + \left(\frac{kV_m \cos \omega_m t}{\omega_m RC}\right)^2} \quad \text{and} \quad \tan \Delta\phi = \frac{kV_m}{\omega_m RC} \cos \omega_m t$$

As $\omega_m RC \gg 1$, $\tan \Delta\phi \simeq \Delta\phi$ and we obtain

$$v_R \simeq V_c \sqrt{1 + \left(\frac{kV_m \cos \omega_m t}{\omega_m RC}\right)^2} \sin \left[\omega_c t - \frac{kV_m}{\omega_m RC} \cos \omega_m t\right]$$

The expression for v_R is similar to that for an FM wave but with a slight amount of AM due to the second term under the square root sign, which can be removed by a limiter. The frequency deviation $\Delta f = kV_m/2\pi RC$ and is initially small since $\Delta\phi$ is about 0·5 radian for good linearity.

4.3 PM/FM transmitter

The initial FM produced by the modified modulating signal is too small and occurs at the lowest modulating frequency of 50 Hz. Since $\Delta\phi$ is usually about 0·5 radian, it amounts to $\Delta f = 50 \times 0{\cdot}5 = 25$ Hz. As typical FM systems use a value of $\Delta f = \pm 75$ kHz, it calls for a frequency multiplication of about 3000 times.

To achieve this, several stages of multiplication are required together with some mixing stages, to ensure that the final carrier is at the appropriate frequency which usually lies in the VHF band. A typical system is shown in

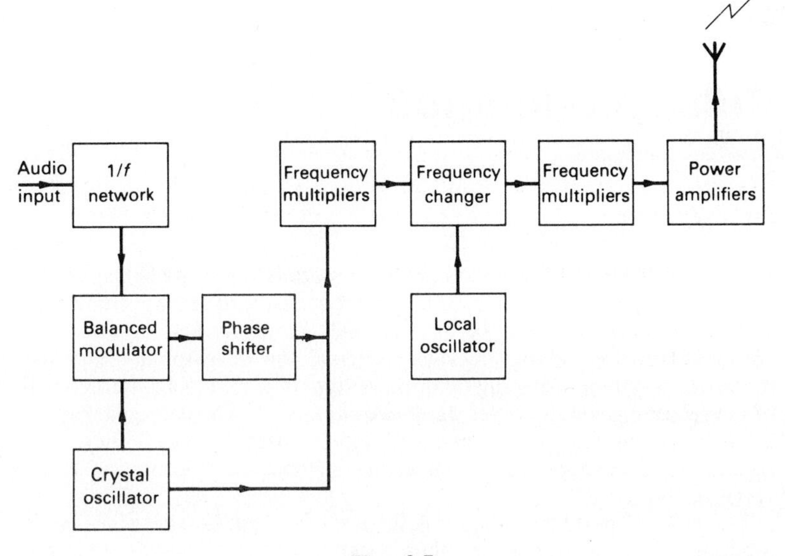

Fig. 4.5

Fig. 4.5. It uses a crystal oscillator with an initial frequency shift of ± 25 Hz and this is multiplied by a set of multipliers to $\Delta f = \pm 1{\cdot}5$ kHz. A mixer stage at 10·8 MHz produces a difference frequency of 1·8 MHz which is subsequently multiplied to 90 MHz $\pm$ 75 kHz. Power amplifiers then produce an output power level of 10 W prior to transmission.

5

Pulse modulation

Up to now, the use of analogue signals for modulating a carrier wave has been considered. Systems using this technique are widespread and will continue to be so for certain applications because of their simplicity and superior characteristics. However, alongside such systems, the development and use of modulation systems using digital or pulse signals has led to an alternative form of modulation generally called *pulse modulation.*[2, 18] The particular properties of such systems need to be examined and evaluated, to see if they might in certain cases have properties superior to those of the existing analogue systems.

The basis of pulse modulation is the use of a digital carrier signal which is modulated by an analogue modulating signal. This may be achieved in various ways and it gives rise to specific types of pulse modulation. These will now be examined in turn and a comparison made as regards techniques and signal-to-noise performance.

5.1 Pulse amplitude modulation

In pulse amplitude modulation (PAM), the amplitude of a train of digital pulses is varied in proportion to the amplitude of the modulating signal. Basically, the modulating signal is 'sampled' by the digital train of pulses and the process is based upon the Sampling theorem.

For simplicity, the analogue signal is assumed to be a single sine wave of frequency f_m and the digital carrier wave is represented by a train of rectangular pulses of 1 V peak amplitude, with a sampling frequency f_s and pulse width τ. This is illustrated in Fig. 5.1(a) and the sampling circuit in Fig. 5.1(b).

The unmodulated pulse train is represented by the expression[18, 19]

$$v_i(t) = \frac{\tau}{T} + \frac{2\tau}{T} \sum_{n=1}^{n=\infty} \frac{\sin(n\omega_s\tau/2)}{(n\omega_s\tau/2)} \times \cos n\omega_s t$$

where the pulses are normalised to 1 V peak and $T = 1/f_s$.

For a modulating signal of the form $v_m = mV \sin \omega_m t$, where $V = 1$ volt and m is the modulation factor (depth of modulation), the modulating process is basically one of amplitude modulation. The PAM train is simply obtained by

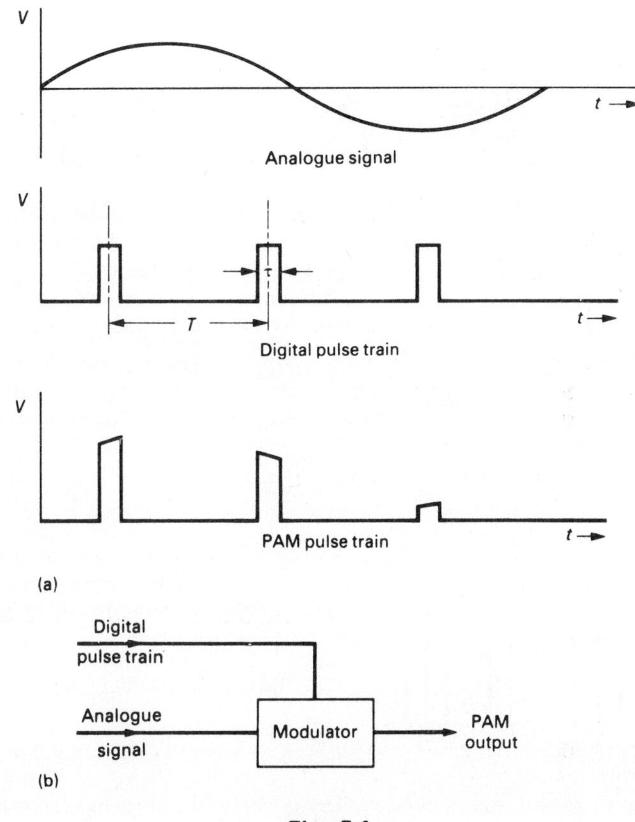

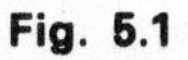
Fig. 5.1

multiplying the carrier voltage $v_i(t)$ by a factor $(1 + m \sin \omega_m t)$ as in AM. Hence, the modulated pulse train is given by

$$v_c(t) = (1 + m \sin \omega_m t)\left[\frac{\tau}{T} + \frac{2\tau}{T}\sum_{n=1}^{n=\infty}\frac{\sin(n\omega_s\tau/2)}{(n\omega_s\tau/2)} \times \cos n\omega_s t\right]$$

or $$v_c(t) = \frac{\tau}{T} + \frac{m\tau}{T}\sin \omega_m t + \frac{2\tau}{T}\sum_{n=1}^{n=\infty}\frac{\sin x}{x}\cos n\omega_s t$$

$$+ \frac{2m\tau}{T}\sum_{n=1}^{n=\infty}\frac{\sin x}{x}\cos n\omega_s t \times \sin \omega_m t$$

where $x = n\omega_s\tau/2$. Hence

$$v_c(t) = \frac{\tau}{T} + \frac{m\tau}{T}\sin\omega_m t + \frac{2\tau}{T}\sum_{n=1}^{n=\infty}\frac{\sin x}{x}\cos n\omega_s t$$

$$+\frac{m\tau}{T}\sum_{n=1}^{n=\infty}\frac{\sin x}{x}[\sin(\omega_s+\omega_m)t + \sin(\omega_s-\omega_m)t]$$

The frequency spectrum of the modulated pulse train may be obtained by considering first that the unmodulated pulse train consists of a set of discrete frequency components at f_s, $2f_s$, $3f_s$, etc. Due to amplitude modulation, each component will have a set of lower sideband and upper sideband components on either side. The modulation spectrum is therefore easily obtained as in Fig. 5.2(a).

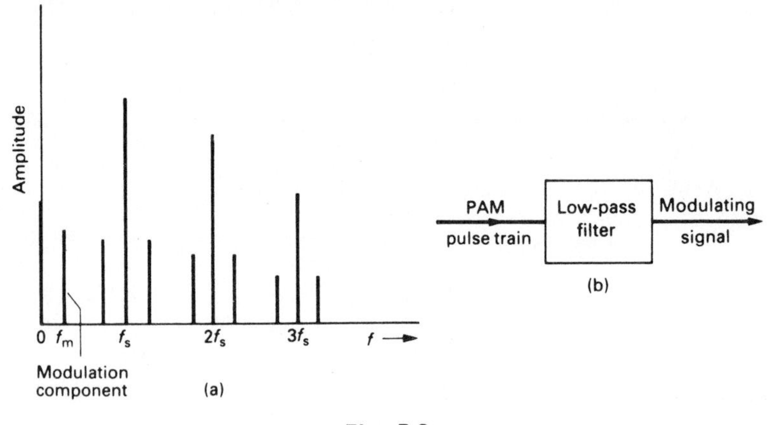

Fig. 5.2

In particular, it is observed from the expression for $v_c(t)$ that the original analogue modulation occurs directly in the output at frequency f_m with amplitude $m\tau/T$. Hence, to recover the modulation, the modulated pulse train is passed through a low-pass filter as shown in Fig. 5.2(b) and the output contains a d.c. component and the modulating signal.

This system can be used for a more complex modulating signal, such as speech. However, the sampling frequency is adjusted to meet the limitation on the bandwidth of the speech, which must therefore be passed first through an appropriate filter before entering the PAM modulator. Furthermore, the PAM technique suffers from the same signal-to-noise limitations as does amplitude modulation. Hence, it is not generally used for a complete system but is largely employed as a basic process in other pulse systems, such as PDM and PPM.

5.2 Pulse duration modulation

The technique of varying the width of a pulse by the modulating signal is called pulse width modulation (PWM) or pulse duration modulation (PDM). Either the leading edge or the lagging edge, or both, may be varied by the modulating signal.

For a system in which the leading edge of the pulse is maintained constant, while the trailing edge is varied in position, assume the leading pulse edges occur at times 0, $1/f_s$, $2/f_s$, etc., where f_s is the sampling frequency of the pulse train. If the modulating signal is of the form $v_m = mV \sin \omega_m t$, where $V = 1$ volt, then the pulse durations vary as $\tau(1 + m \sin \omega_m t)$, where τ is the unmodulated pulse duration or width.

Initially, let the pulse train be $v_i(t)$ which is given by

$$v_i(t) = \frac{\tau}{T} + \frac{2\tau}{T} \sum_{n=1}^{n=\infty} \frac{\sin x}{x} \cos n\omega_s t$$

where $x = n\omega_s \tau/2$ and the pulses are normalised to a peak value of 1 volt. Hence, we obtain the modulated pulse train as $v_c(t)$ with

$$v_c(t) = \frac{\tau}{T}(1 + m \sin \omega_m t) + \sum_{n=1}^{n=\infty} \frac{2}{n\pi} \sin\left[\frac{n\omega_s \tau}{2}(1 + m \sin \omega_m t)\right] \cos n\omega_s t$$

or

$$v_c(t) = k + mk \sin \omega_m t + \sum_{n=1}^{n=\infty} \frac{2}{n\pi} \sin\left[\frac{n\omega_s \tau}{2}(1 + m \sin \omega_m t)\right] \cos n\omega_s t$$

where $k = \tau/T = \tau f_s$ is a constant.

The first term is a d.c. component which may be blocked by a capacitor, while the second term corresponds to the modulating signal multiplied by a factor mk. If the other side-frequencies in the expression are sufficiently far away from f_m, the modulation can be recovered by passing the modulated signal through a low-pass filter. The technique is shown in Fig. 5.3(a), where the output pulse width varies with the modulation. A typical demodulator is shown in Fig. 5.3(b).

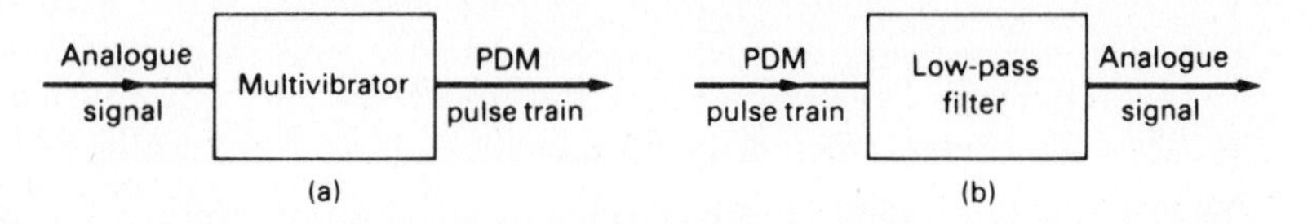

Fig. 5.3

PDM gives a better signal-to-noise performance than PAM as noise affects mainly the trailing edge which carries the modulation. Noise at the leading edge or during the pulse interval has a smaller effect than in PAM. The overall noise effect may be minimised by using steeper trailing edges. This result follows directly from the fact that PDM uses a larger bandwidth to achieve the better signal-to-noise ratio. In Appendix D it is shown that $v_c(t)$ when expanded represents a PM wave and so it behaves similarly as regards signal-to-noise performance.

5.3 Pulse position modulation

The process of varying the position or time occurrence of a pulse due to the modulating signal is called pulse time modulation (PTM) or pulse position modulation (PPM). This is achieved by shifting each pulse from its unmodulated position by an amount proportional to the modulating signal.

If the unmodulated pulse train is represented by $v_i(t)$, we have

$$v_i(t) = \frac{\tau}{T} + \frac{2\tau}{T} \sum_{n=1}^{n=\infty} \frac{\sin(n\omega_s\tau/2)}{n\omega_s\tau/2} \cos n\omega_s t$$

or

$$v_i(t) = \tau f_s + \sum_{n=1}^{n=\infty} \frac{2}{n\pi} \sin(n\pi f_s \tau) \cos(2n\pi f_s t)$$

if the sampling frequency is f_s, where $f_s = 1/T$ and the centre of each pulse occurs at the time instants 0, T, $2T$, etc. Due to the modulating signal, $v = \sin\omega_m t$, the centre of each pulse is displaced in time by an amount $\Delta t \sin\omega_m t$ or $mT \sin\omega_m t$, where $m = \Delta t/T$. Hence, the modulated pulse train is given by

$$v_c(t) \simeq \frac{\tau}{T} + m\omega_m \tau \cos\omega_m t$$

$$+ \frac{2\tau}{T} \sum_{n=1}^{n=\infty} \frac{\sin x}{x} (1 + m\omega_m T \cos\omega_m t) \cos n[\omega_s t + \phi(t)]$$

where $x = n\omega_s\tau/2$ and $\phi(t) = m\omega_s T \sin\omega_m t$. (See Appendix D.)

Note
The third term can be expanded further into an infinite series of Bessel functions.

In Appendix D it is shown that $v_c(t)$ contains a modulation term and a set of phase-modulated waves. The modulation factor m is generally small but the

PPM output yields a somewhat better signal-to-noise ratio than do those of PAM or PDM. This is because noise has a smaller disturbing effect on the time position of a pulse edge which carries the modulation.

To recover the modulation, the PPM pulses are usually converted back to PDM or PAM pulses at the receiver and then passed through a low-pass filter. The technique is shown in Fig. 5.4.

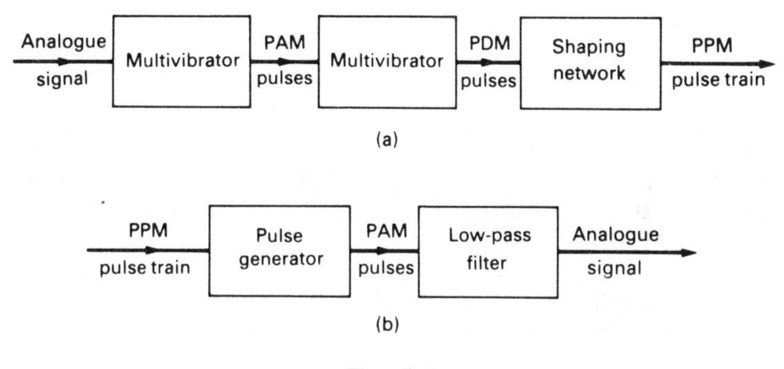

Fig. 5.4

5.4 Pulse code modulation[20]

A form of modulation which uses coded groups of pulses to represent certain values of the modulating signal is called pulse code modulation (PCM). Since the modulating signal consists of a continuous set of values, it is divided into a *finite* set of discrete values between an upper and lower limit—a technique known as *quantisation.* Such a quantised signal is an approximation to the analogue signal and, in uniform quantisation, the discrete levels are equally spaced while, in non-uniform quantisation, they are unequally spaced. This is illustrated in Fig. 5.5.

The analogue signal to be transmitted is first sampled and the 'samples', which correspond to a set of PAM pulses, are then rounded off to the nearest quantised level in the quantiser. The quantised pulses are usually coded into groups according to the binary code. Each pulse group represents its quantised level as a binary number and the maximum number of pulses in a group depends upon the total number of quantised levels chosen for the system. Usually 128 levels are used for speech signals, where $2^7 = 128$, and so a 7-pulse grouping is employed with non-uniform quantisation.

The differences between the analogue signal levels and the quantised signal levels lead to an *uncertainty* which is known as *quantisation noise.* This type of noise can only be reduced by using a greater number of levels, e.g. 256 levels would require an 8-pulse group, but the use of a greater number of levels is

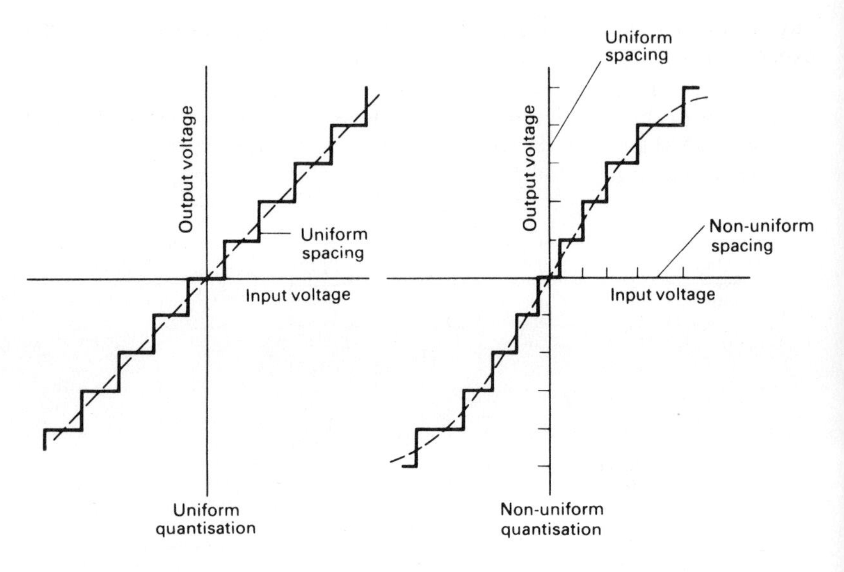

Fig. 5.5

expensive in bandwidth. In general, if q is the number of quantised levels, p is the maximum number of pulses in a group, and l is the number of voltage levels of the pulse, then we have

$$l^p = q$$

or

$$p = \log_l q \text{ pulses}$$

and for binary PCM, which uses two voltage levels of 0 V and +1 V, we have $l = 2$ and so

$$p = \log_2 q \text{ pulses}$$

Some power economy is obtained by using a bipolar, binary code with voltage levels of +1 V and −1 V, as no d.c. level need be transmitted because it conveys no information. Furthermore, some signal-to-quantisation noise improvement is possible without increasing the number of levels by using non-uniform quantisation. This amounts to using closer levels at low signal values and wider levels at large signal values. Less noise is produced at the lower levels, while the greater amount of noise produced at the higher levels appears less objectionable due to the large signal values.

In practice, non-uniform quantisation is achieved by compressing the speech signal at the transmitter using a non-linear *compressor* and then quantising the compressed signal in a linear quantiser. At the receiver, the signal is restored to its normal levels by using an *expander* with the opposite characteristic to cancel any signal distortion produced. The combination of a compressor and an expander is known as a *compander*.

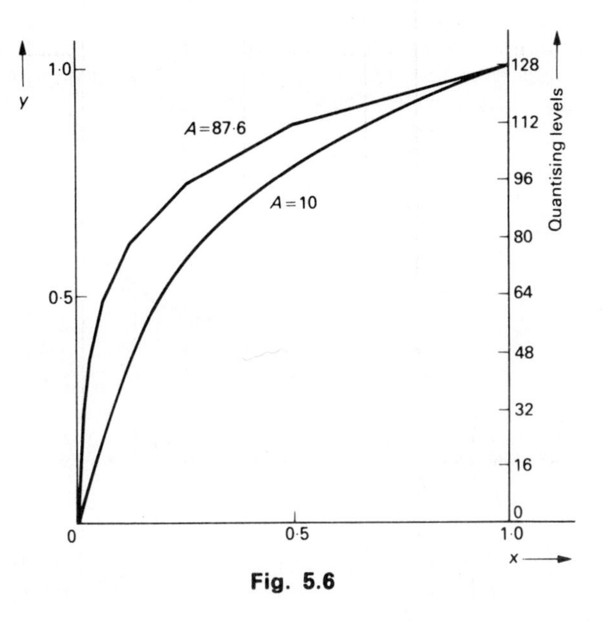

Fig. 5.6

A typical transfer characteristic of the compressor is shown in Fig. 5.6 and it uses a μ-law given by [21]

$$|v_o| = \frac{\log(1+\mu|v_i|)}{\log(1+\mu)}$$

where v_i and v_o are the *normalised* input and output voltages respectively and μ is a parameter which yields linear quantisation when $\mu = 0$.

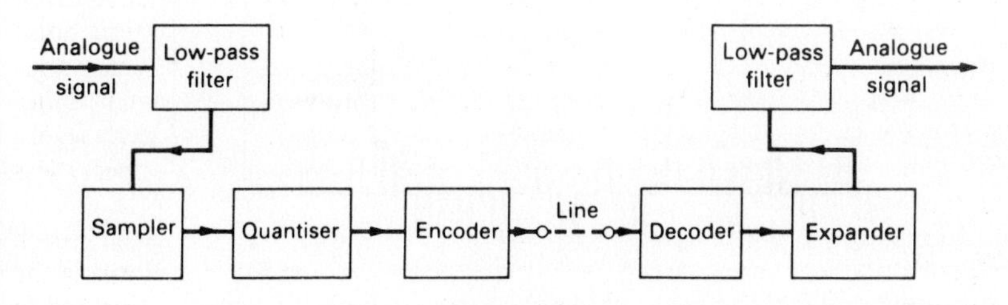

Fig. 5.7

A block schematic arrangement of a single-channel PCM system is shown in Fig. 5.7. The analogue signal is first band-limited to a maximum frequency W by a low-pass filter in accordance with the Sampling theorem and it is then sampled. The sampled pulses are rounded off to the nearest quantised level in the quantiser and the quantised samples are converted into groups according to the binary code in the encoder. The appropriate signals are shown in Fig. 5.8.

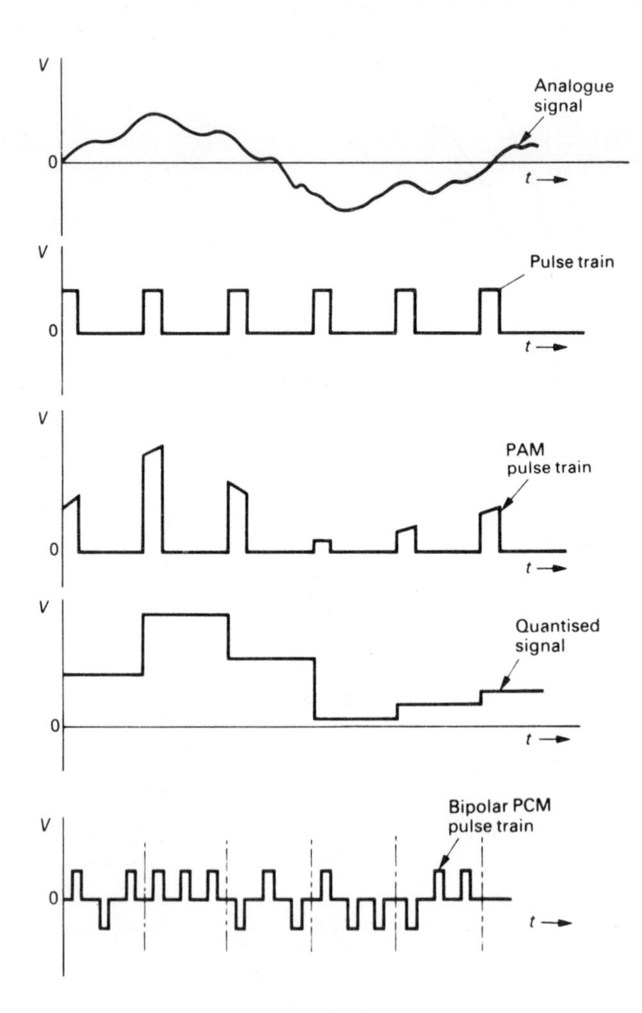

Fig. 5.8

The transmitted pulse train in a bipolar PCM system consists of the pulse group combinations where a positive pulse signifies '1' and a negative pulse '0'. It is transmitted to the receiving end and, at the receiver, the pulse groups are decoded in the decoder. This results in a train of individual AM pulses, each pulse representing a sampled level which was later quantised. After passing the pulse train through a low-pass filter, the original analogue signal is recovered but with added circuit noise and quantisation noise.

Example 5.1

Explain how pulse code modulation (PCM) is used to transmit a continuous signal waveform as a train of binary digits. What are the advantages of PCM?

Describe, with the aid of a block diagram, a time division multiplex system using PCM to transmit four telephone signals along one physical circuit.

Estimate the number of bits per second required if each signal contains frequencies between 0 and 3 kHz and is quantised into 64 levels. (C.E.I)

Solution

The answer to the first part of the question is given at the beginning of this chapter. The advantages of PCM are mentioned in Example 5.2.

TDM system

Since PCM requires a large bandwidth for transmission, some economy in bandwidth is possible by multiplexing several channels on a time division basis. A schematic arrangement for four telephone channels is shown in Fig. 5.9.

In Fig. 5.9, the speech signals, after filtering, are fed into channel gates which are opened in turn by a time sequence pulse train. These pulses sample the speech signals sequentially and are then quantised. The four outputs are combined and, after encoding, are transmitted along the same line. At the receiving end, the time-multiplexed pulse train is decoded and enters channel gates which are opened sequentially to correspond with those at the transmitting end. The PCM pulses separate into the four channels and the speech signals are recovered by filtering.

Problem

For a bandwidth of 0 to 3 kHz, the sampling frequency should be at least twice this and it is usually taken as 8 kHz for CCITT voice channels of 4 kHz which includes a guard space. To accommodate four channels, this must be increased four times. Hence, $f_s = 4 \times 8 \times 10^3 = 32\,\text{kHz}$.

For 64 levels, we require 2^6 or a 6-pulse group for a PCM system. The pulse rate is then equal to $6 \times 32 \times 10^3$ or 192 kilobit/s. If a timing pulse is added to each group for synchronisation, we obtain a pulse rate of $7 \times 32 \times 10^3$ or 224 kilobit/s.

5.5 TDM/PCM telephone system

A typical PCM system for junction telephone circuits between local exchanges employs audio cables and time-multiplexing of 24 speech channels. It uses a 1·536 Mbit/s digital signal with an 8-pulse group and employs regenerative repeaters.

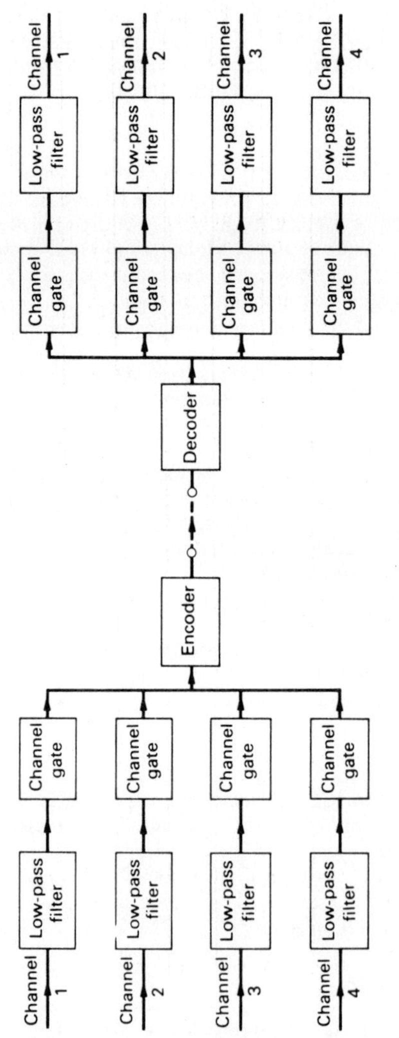
Channel 1
Channel 2
Channel 3
Channel 4
Low-pass filter
Low-pass filter
Low-pass filter
Low-pass filter
Channel gate
Channel gate
Channel gate
Channel gate
Encoder
Decoder
Channel gate
Channel gate
Channel gate
Channel gate
Low-pass filter
Low-pass filter
Low-pass filter
Low-pass filter
Channel 1
Channel 2
Channel 3
Channel 4

Fig. 5.9

Signals are sampled using 8 kHz as the sampling frequency. With a 7-pulse group, 128 levels are used with non-uniform (logarithmic) quantisation. An eighth pulse is added to each group for time synchronisation between transmitter and receiver. The repeaters are spaced every 2 km apart and use semiconductor integrated circuits. Details of the system are given in Fig. 5.10. An alternative system uses 32 timeslots for 30 speech channels and operates at 2·048 Mbit/s.

5.6 Main features of PCM

The increasing use of PCM for information transmission is finding greater application in both short-distance and long-distance communication. The main advantage of PCM is its very good signal-to-noise ratio obtained at the expense of an increased bandwidth. Since it is only necessary to detect the presence or absence of a pulse, pulses can be regenerated at regular intervals along a line, thus ensuring no degeneration of S/N ratio along the route. This is in marked contrast with other systems in which the S/N ratio gets progressively worse through the system. Hence, one important future application of PCM is in trunk telephony (30 km or more), through the use of repeater stations employing pulse regeneration.

However, the main disadvantages of PCM are its large bandwidth and complexity. The increased bandwidth is due to the fact that $2nW$ pulses have to be transmitted per second, where n is the number in each PCM pulse group and W is the highest frequency component transmitted. The complexity is due to the fact that PCM requires the regeneration and encoding of narrow pulses. It also requires a precision timing system to minimise digital errors. Consequently, the use of high-speed circuits, such as encoders, is a problem which is being solved by developments in integrated circuit technology.

5.7 PCM noise[22]

The two important sources of noise are transmission noise and quantisation noise. Transmission noise occurs all along the channel and is the familiar white Gaussian noise due to thermal sources. It gives rise to *bit errors* in the transmitted PCM pulses by changing a 'zero' into a 'one' or vice versa. For unipolar PCM pulses varying between zero level and peak amplitude A, the probability of error P_e is similar to that obtained in a digital data system using *amplitude-shift-keying* (ASK).

It can be shown that the probability of error P_e is given by

$$P_e = \frac{1}{2}\operatorname{erfc}\left(\sqrt{\frac{C}{4N}}\right)$$

or

$$P_e = \frac{1}{2}\operatorname{erfc}\left(\frac{1}{2\sqrt{2}}\frac{A}{\sigma}\right)$$

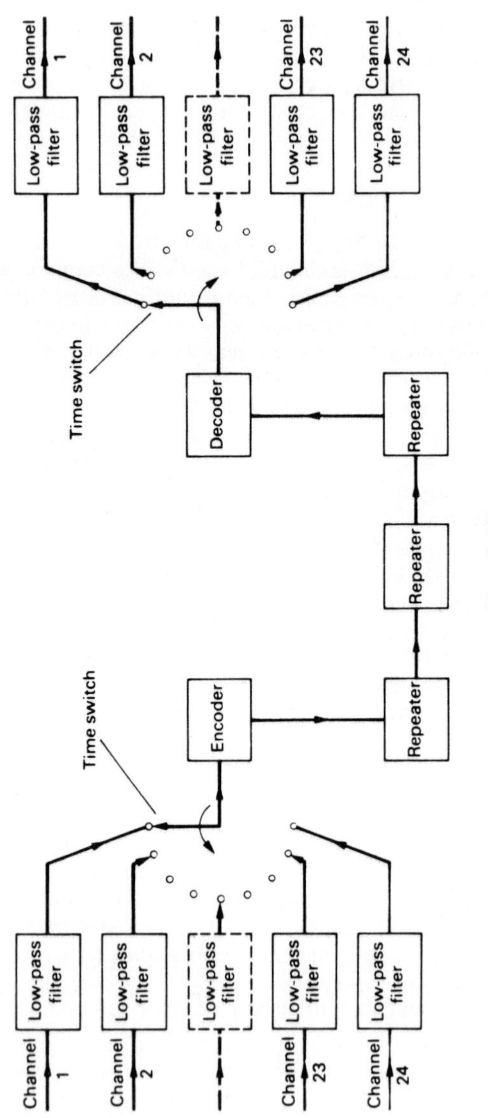
Channel 1
Channel 2
Channel 23
Channel 24
Low-pass filter
Low-pass filter
Low-pass filter
Low-pass filter
Low-pass filter
Time switch
Encoder
Repeater
Repeater
Repeater
Decoder
Time switch
Low-pass filter
Low-pass filter
Low-pass filter
Low-pass filter
Low-pass filter
Channel 1
Channel 2
Channel 23
Channel 24

Fig. 5.10

where erfc is the complimentary error function and $(C/N) = A^2/2\sigma^2$ is the carrier-to-noise power ratio, if A is the peak pulse amplitude and σ is the rms noise voltage. Various values of P_e and the ratio (A/σ)dB are shown in Table 5.1.

Table 5.1

P_e	(A/σ)dB
10^{-4}	17·4
10^{-6}	19·6
10^{-8}	21·0
10^{-10}	22·0

For input (S/N) ratios of about 21 dB, which is called the error threshold, transmission noise effects can be ignored. When operating above this threshold, quantisation noise is mainly important. It arises due to the 'uncertainty' in transmitting a particular signal level caused by rounding-off (up or down) in the quantiser and can only be reduced by using more levels or non-uniform quantisation. It is given by the expression

$$(S_o/N_o) \simeq 2^{2n}$$

or

$$(S_o/N_o) \simeq 2^{2(B_c/B)}$$

where the exponent $n = B_c/B$, if B_c is the channel bandwidth and B is the baseband bandwidth. Typical values for various numbers of quantising levels $L = 2^n$ are given in Table 5.2.

Table 5.2

L	(S_o/N_o)dB
32	32
64	38
128	44
256	50

The previous expression for the output signal-to-quantisation noise ratio (S_o/N_o) shows that it increases exponentially with the channel bandwidth B_c used because of the exponent $n = B_c/B$. This is usually true for practical values of (S_o/N_o) greater than about 10.

Since this is also true for an ideal system as proposed by Shannon, comparison with an ideal system indicates that the power requirements of PCM for similar conditions of minimum error transmission are about 8 dB greater than those of the ideal, as illustrated in Fig. 5.11.

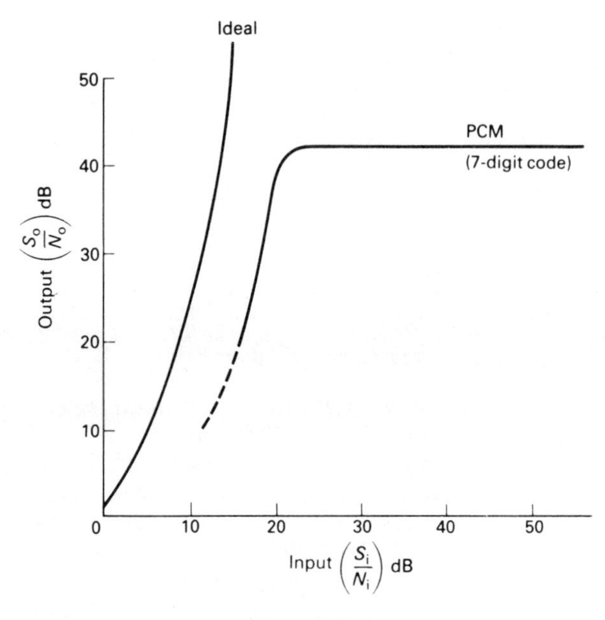

Fig. 5.11

Example 5.2

Discuss the particular advantages that pulse code modulation has over other methods for use in a large telecommunication system. Are there any major disadvantages?

Derive a formula for the channel capacity of a PCM system in which the signal has a bandwidth f, there are M quantising pulses, and a code group consists of m pulses of l levels each.

Discuss briefly how a PCM system compares with Shannon's ideal system.

(C.E.I.)

Solution

The particular advantages of PCM over other systems are:

1. It has superior S/N characteristics for a given bandwidth.
2. It is well suited to long-distance communication without loss of S/N ratio due to pulse regeneration.
3. It fits in with other digital systems, such as data transmission.

The major disadvantages of PCM are:

1. It requires a very large bandwidth.
2. It uses costly, sophisticated circuits, which have to be duplicated several times if many repeaters are used.
3. It is generally uneconomical for short distances, i.e. < 15 km. This limitation is being overcome by the use of integrated circuits.

Problem
The communication capacity of a system is given by

$$C = 2W \log_2 (1 + S/N)^{1/2} = 2W \log_2 n \text{ bits/s}$$

where W is the bandwidth of the system and n is the number of distinguishable voltage levels.

For the PCM system, we have $W = f$ and $l^m = M$, hence

$$C = 2f \log_2 M = 2mf \log_2 l \text{ bits/s}$$

Final part
The PCM system is the nearest approach to Shannon's ideal system. However, it is still about 8 dB below it which implies that to obtain the same error rate as predicted for the ideal system, a PCM system needs about 6 times as much power as the ideal system. In other words, it is only about 17% as efficient as the ideal system.

5.8 Delta modulation[23]

Delta modulation (DM) uses a *one*-digit code which conveys information about the *derivative* of the signal amplitude instead of the actual amplitude as in PCM. This is done by integrating the output from the modulator and then comparing it with the input signal to the comparator. A pulse is then transmitted and its polarity is such as to reduce the *difference* signal into the modulator.

At the receiver, the transmitted pulses are integrated and passed through a low-pass filter to remove unwanted high-frequency components. The output consists of the original analogue signal together with some additional noise somewhat similar to quantisation noise. A block schematic arrangement is shown in Fig. 5.12.

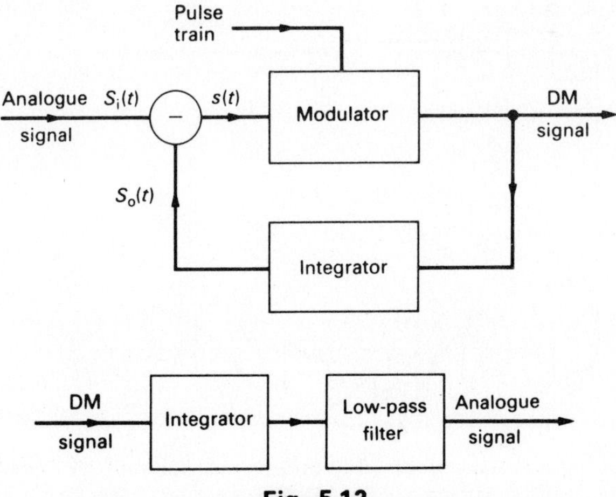

Fig. 5.12

The analogue input signal $S_i(t)$ and the integrated output signal $S_o(t)$ are compared in the comparator and the difference signal $s(t) = S_i(t) - S_o(t)$ enters the modulator, together with a train of clock pulses from a pulse generator. When $s(t)$ is positive, a positive pulse is transmitted and, when $s(t)$ is negative, a negative pulse is transmitted. If $s(t)$ is zero, then alternate positive and negative pulses are transmitted as shown in Fig. 5.13 (a).

The receiver consists of an integrator followed by a low-pass filter. The received signal on integration yields a step-like waveform which closely follows the original analogue signal as shown in Fig. 5.13 (b) and any difference between them appears as 'quantised noise'. This is shown in Fig. 5.13(c).

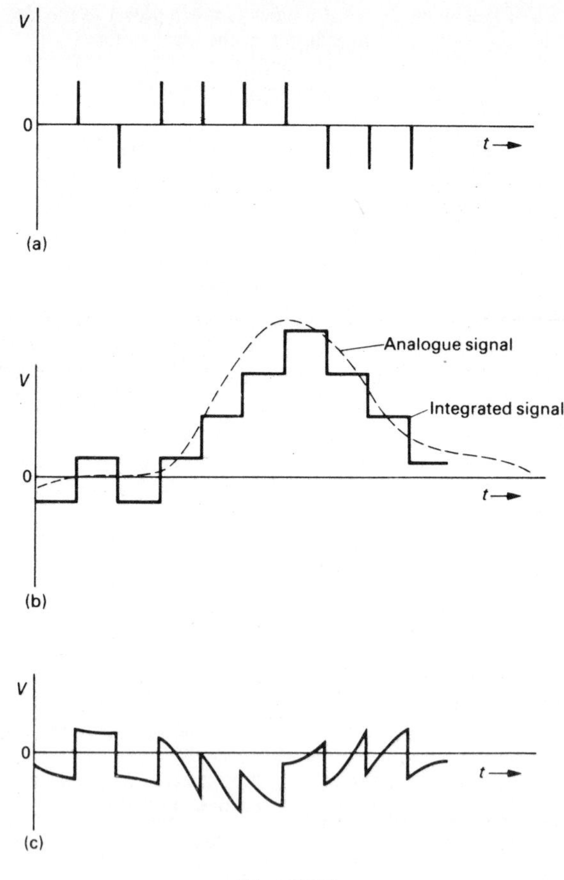

Fig. 5.13

The condition for proper transmission is obtained when

$$\frac{d}{dt}[S_i(t)] \times T_s \leqslant \sigma$$

where $S_i(t)$ is the input signal, T_s is the sampling period, and σ is the magnitude of the voltage 'steps'. For a sinusoidal input of frequency f_m, we obtain

$$S_i(t) = V_m \sin \omega_m t$$

and

$$\frac{d}{dt}[S_i(t)] = \omega_m V_m \cos \omega_m t$$

Hence

$$\omega_m V_m \cos \omega_m t \times T_s \leqslant \sigma$$

or

$$\omega_m V_m \times T_s \leqslant \sigma$$

with

$$\frac{V_m}{\sigma} \leqslant \frac{f_s}{2\pi f_m}$$

where $f_s = 1/T_s$ is the sampling frequency.

Since $2V_m/\sigma$ is the number of voltage steps or levels in the peak-to-peak analogue signal, we obtain

$$\text{number of steps} \leqslant \frac{f_s}{\pi f_m}$$

If the highest modulating frequency is W, the minimum number of steps is at least one and so $f_s \geqslant \pi W$. This result is therefore higher than the value $2W$ obtained by the Sampling theorem for PCM.

The main application of DM is for speech transmission and, to avoid overload slope distortion or to reduce quantisation noise, the sampling frequency must be high. It can be shown that the output signal-to-quantisation noise in single integration DM and based upon sine-wave considerations is given by

$$\left(\frac{S}{N_q}\right)_o \simeq \frac{0{\cdot}02 f_s^3}{B f^2}$$

where f_s is the sampling frequency, B is the baseband signal bandwidth, and f is the input signal frequency. Hence, the ratio is small at the higher frequencies unless f_s is large. However, it can be shown that DM at a 40 kbit/s pulse rate achieves the same performance as a 5-bit code PCM system.

A variation of DM which achieves better performance is *adaptive delta modulation* (ADM).[24] It uses a variable quantum 'step' size which is small for a slowly changing waveform and increases to a larger value as the waveform changes steeply. The adaptive technique produces small quantisation and overload errors and a reasonable telephone transmission can be achieved using a bit stream of 32 kbit/s instead of the usual 64 kbit/s required by a PCM system.

Example 5.3

Discuss the difference between delta modulation and pulse code modulation comparing the merits and demerits of the two systems.

Explain the meaning of the two terms 'quantisation' and 'quantisation noise'.

By means of a diagram show how the quantised signal is related to these two signals.

Draw a block diagram of a delta modulator. (C.G.L.I.)

Solution

The main differences between delta modulation and pulse code modulation are:

1. Delta modulation transmits information about the *difference* between adjacent signal levels sampled, whereas pulse code modulation conveys information about the *actual* levels sampled.
2. Delta modulation uses a 1-pulse code for speech signals, whereas PCM normally uses a 7-pulse code for similar signals.

Merits

1. Delta modulation is much simpler than PCM and it simplifies both transmitting and receiving equipment.
2. Delta modulation is almost as good as PCM in performance.

Demerits

1. Delta modulation generally requires a larger bandwidth than PCM for comparable performance.
2. Delta modulation is not ideal for video signals since a d.c. level cannot be transmitted directly by DM.
3. S/N ratio in DM is a function of signal frequency, the higher the frequency, the poorer is the ratio.

Quantisation

Before a signal is transmitted, it is divided into a set of discrete levels which may be as many as 128 for a 7-pulse code. This process is called quantisation and the sampled pulses can only have values corresponding to these levels and not in between. This amounts to approximating a continuous signal by a stepped waveform, as shown in Fig. 5.13(b).

Quantisation noise

The difference in level between the actual signal and the stepped or quantised signal leads to an *uncertainty* which has an effect similar to noise and is called quantisation noise. It may be reduced only by using a greater number of quantised levels as this reduces the interval between voltage levels. It is shown in Fig. 5.13(c).

Delta modulator

A block schematic diagram of a delta modulator is given in Fig. 5.12.

5.9 Delta sigma modulation[25]

The main drawbacks of delta modulation are its inability to transmit d.c. signals and the variation in S/N ratio with frequency. These are overcome by delta sigma modulation (DSM) which uses an integrator at the input to the modulator. In this way, the transmitted pulses convey information about the actual signal amplitude and not just its difference as in DM. Furthermore, by using only one integrator at the transmitter, the integrator at the receiver can be dispensed with, and only a low-pass filter is required at the receiving end. A block schematic arrangement is shown in Fig. 5.14.

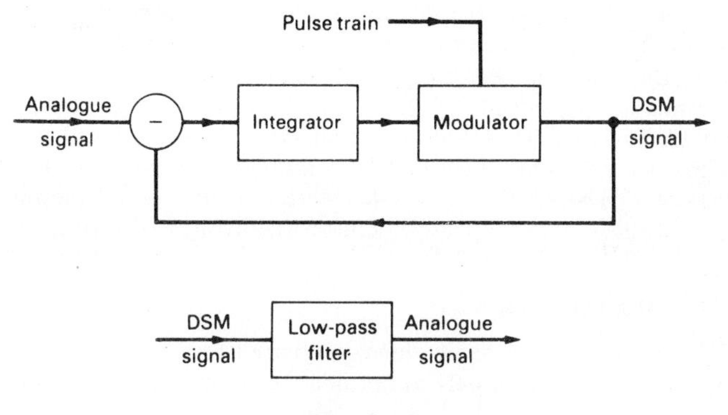

Fig. 5.14

A closer analysis of DSM reveals that apart from its ability to transmit d.c. signals, which is important in video systems, the output signal-to-noise ratio is independent of the modulating signal frequency and so also is the dynamic range of the input signal. Moreover, as no integrator is used at the receiver, errors due to noise disturbances during transmission are not cumulative as in delta modulation.

5.10 Differential PCM[26,27]

Unlike delta modulation which codes pulse differences using a 1-bit code, differential PCM (DPCM) codes pulse differences using normal PCM coding. The technique can be applied to speech or video signals but is being used largely with colour television. For broadcast quality video signals, digital encoding by PCM requires very high bit rates. For example, a PAL 625-line signal with 8-bits per sample, requires a bit rate of about 100–120 Mbit/s. Hence, to reduce costs, it is important to reduce the bit rate without impairing picture quality.

In a video signal, there is a fairly high correlation between samples corresponding to closely adjacent parts of the picture. DPCM exploits the redundancies by using the differences between each sample of the video signal and a predicted value based on previous-sample values. For monochrome television, the previously-transmitted sample serves as the predicted sample for the following sample. By arranging that the encoder and the decoder both use the same prediction of each sample, the original samples are recovered by adding the transmitted differences to the predicted sample values at the receiver.

However, with previous-sample prediction, any reduction in the bit rate leads to larger quantising errors in high-amplitude, high-frequency components of the video signal. Hence, one method of accurately coding the PAL colour signal is to use a sampling frequency which is three times the colour subcarrier frequency of 4·43 MHz and taking differences between every third sample. The presence of the colour subcarrier has little effect on the size of the difference signal, but the colour information is still accurately encoded.

The results of subjective tests on the picture impairment caused by DPCM coding of the PAL 625-line video signal indicate that, for a given picture quality, a DPCM system using third-previous sample prediction at a sampling frequency of 13·3 MHz requires about two fewer bits per sample than PCM coding. A block schematic arrangement of the layout is shown in Fig. 5.15.

5.11 Digital modulation

Digital data consisting of the two binary signals 'mark' and 'space', or 1 and 0, can be transmitted by varying the amplitude, frequency, or phase of a sinusoidal carrier wave. The three methods are known as amplitude-shift-keying (ASK), frequency-shift-keying (FSK), and phase-shift-keying (PSK) respectively.

However, due to the presence of Gaussian noise, for every symbol transmitted, the receiver must make a choice between the two symbols and so the probability of an error occurring is a useful criterion for comparing various types of digital modulation systems. It is shown in another volume of this series* that the probability of error or bit error rate (BER) is given by

$$P_e = \frac{1}{2}\operatorname{erfc}\left[\frac{E(1-\rho)}{2N_o}\right]^{1/2}$$

where erfc signifies the complimentary error function, E is the energy per bit transmitted, ρ is the correlation coefficient, and N_o is the noise power spectral density.

Amplitude-shift-keying (ASK)

A carrier wave is switched ON and OFF by the binary signals. During a 'mark' a carrier wave is transmitted and during a 'space' it is suppressed. Hence, it is also known as ON–OFF keying and was originally used in multichannel telegraph systems. The signal waveforms transmitted are

$$s_1(t) = A\sin\omega t \qquad \text{(for symbol 1)}$$
$$s_0(t) = 0 \qquad \text{(for symbol 0)}$$

* See F. R. Connor, *Noise* (Appendix L), Edward Arnold (1982).

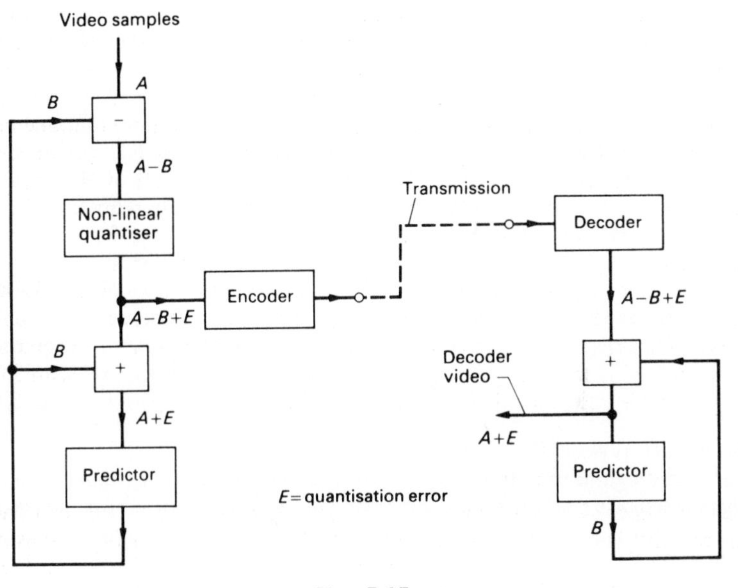

Fig. 5.15

and there is no correlation between the signals, i.e. $\rho = 0$, because the energy during one bit is zero. Hence, we obtain

$$P_e = \tfrac{1}{2}\,\mathrm{erfc}\,(E/4N_o)^{1/2}$$

if the frequency and phase of the transmitted signal are known at the receiver (coherent ASK).

To recover the binary modulation, either coherent synchronous detection or non-coherent envelope detection may be used. Synchronous detection leads to fewer errors than does envelope detection, but it requires the use of a phase-coherent local oscillator at the receiver.

Frequency-shift-keying(FSK)

In this method, two different carrier frequencies are used and they are switched on and off by the binary signals. A mark switches one carrier on while the other carrier is off and a space switches the second carrier on while the first carrier is off. Hence, this amounts to a *form* of ASK in which two *different* carrier frequencies are used.

If the two different frequencies are f_1 and f_0, we have

$$s_1(t) = A \sin \omega_1 t \qquad \text{(for symbol 1)}$$

$$s_0(t) = A \sin \omega_0 t \qquad \text{(for symbol 0)}$$

and detection is by means of two matched filters.[28] For well-separated frequencies, the two signals are orthogonal, i.e. $\rho = 0$, and we obtain

$$P_e = \tfrac{1}{2}\operatorname{erfc}(E/2N_o)^{1/2}$$

a result which is less than the previous case if the frequency and phase of each signal are known at the receiver. The main disadvantage of ASK over FSK is the need for automatic gain control to overcome fading effects at the receiver. However, the lower probability of errors due to noise in FSK is obtained at the expense of a greater bandwidth.

Detection can be achieved by using synchronous or envelope detectors. In the latter case, two band-pass circuits or matched filters are used which are tuned to the two different frequencies respectively. This is non-coherent FSK which leads to some degradation in performance and the error probability is given by

$$P_e = \tfrac{1}{2}e^{-E/2N_o}$$

Phase-shift-keying (PSK)

The binary signals are used to switch the phase of a carrier wave between two values which are usually 0° and 180°. For a mark the carrier has one phase and for a space it is reversed by 180°. Hence, it is sometimes called phase-reversal-keying (PRK).

The waveforms transmitted are

$$s_1(t) = A\sin\omega t \qquad \text{(for symbol 1)}$$
$$s_0(t) = -A\sin\omega t \qquad \text{(for symbol 0)}$$

and the two signals are identical but of opposite phase, i.e. $\rho = -1$, and we obtain

$$P_e = \tfrac{1}{2}\operatorname{erfc}(E/N_o)^{1/2}$$

which is the minimum value of P_e obtainable for a given E/N_o if the frequency and phase are known at the receiver. Hence, this coherent PSK system is the optimum digital modulation system.

To detect the binary signals, coherent detection must be used to detect the phase of the received signal. This requires, as a phase reference, a coherent local oscillator at the receiver which produces synchronising difficulties in practice. To overcome this, a form of PSK which uses, as a reference, the phase information of the previous bit transmitted, and is called differential phase-shift-keying (DPSK), may be employed. It can be shown that the probability of error is given by

$$P_e = \tfrac{1}{2}e^{-E/N_o}$$

which is about 2 dB more than optimum PSK. Furthermore, as the reference bit can be contaminated by noise, digital errors tend to occur in pairs and suitable error-correcting codes must be used to correct them.

To conserve power or bandwidth, multi-amplitude, multi-frequency, or multi-phase signalling may be employed. Of these, quadrature phase-shift-keying (QPSK) is often employed and uses the four phase values 45°, 135°, 225°, and 315°. Its performance is comparable to that of PSK for large values of E/N_0, and it requires only half the bandwidth of the PSK system. Typical waveforms for ASK, FSK, and PSK are shown in Fig. 5.16 and suitable block diagrams are illustrated in Fig. 5.17.

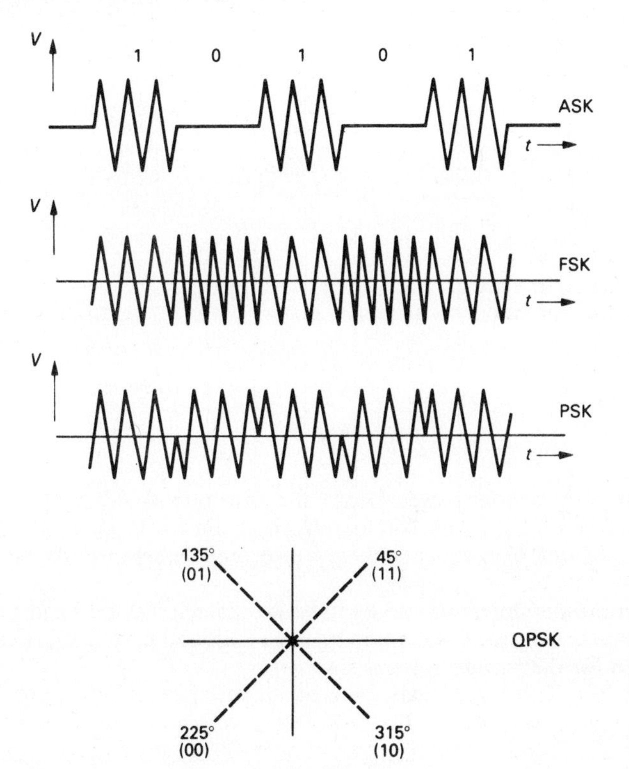

Fig. 5.16

Bit error rate (BER)

It has been shown that the probability of error P_e or bit error rate (BER) due to Gaussian noise depends essentially on the ratio E/N_0. The bit error rate is of primary importance in digital data systems and is typically about 10^{-5} or less.

To reduce errors in a bit stream, the encoded binary data is transmitted with additional *check* digits. This increases the *bit rate* and so reduces the ratio E/N_0

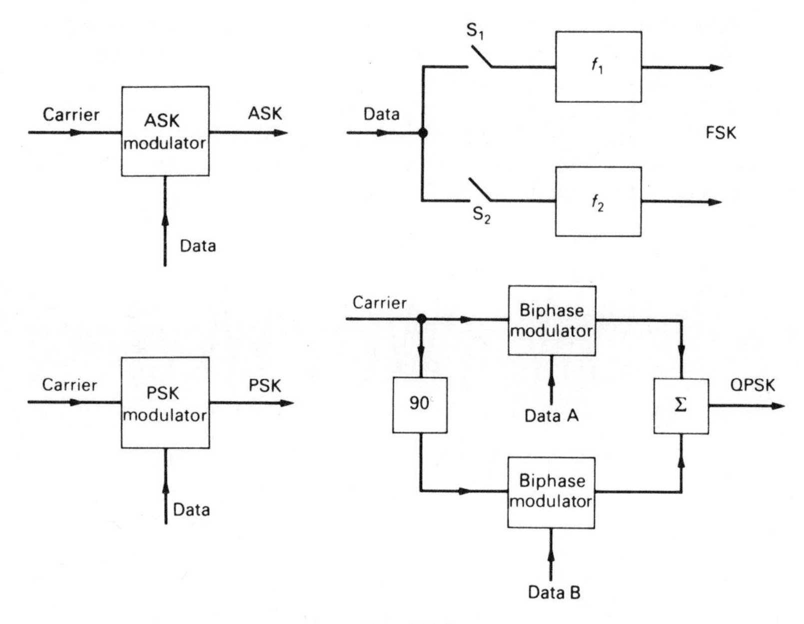

Fig. 5.17

for a given transmitter power. Hence, the error rate would paradoxically have *increased* were it not for the fact that the check digits also correct the additional errors produced. Consequently there is an *overall* reduction of the bit error rate from 10^{-5} to 10^{-7} in a typical case.

A summary of the various values of BER is given in Table 5.3 and a typical set of graphs showing the relationship between BER and E/N_o(dB) is illustrated in Fig. 5.18 for the various systems.

Table 5.3

System	*Type*	*BER*
ASK	Coherent	$\frac{1}{2}\,\mathrm{erfc}(E/4N_o)^{1/2}$
FSK	Coherent	$\frac{1}{2}\,\mathrm{erfc}(E/2N_o)^{1/2}$
PSK	Coherent	$\frac{1}{2}\,\mathrm{erfc}(E/N_o)^{1/2}$
DPSK	Coherent	$\frac{1}{2}\,e^{-(E/N_o)}$
QPSK	Coherent	$\simeq \frac{1}{2}\,\mathrm{erfc}(E/N_o)^{1/2}$
ASK	Non-coherent	$\simeq \frac{1}{2}\,e^{-(E/4N_o)}$
FSK	Non-coherent	$\frac{1}{2}\,e^{-(E/2N_o)}$

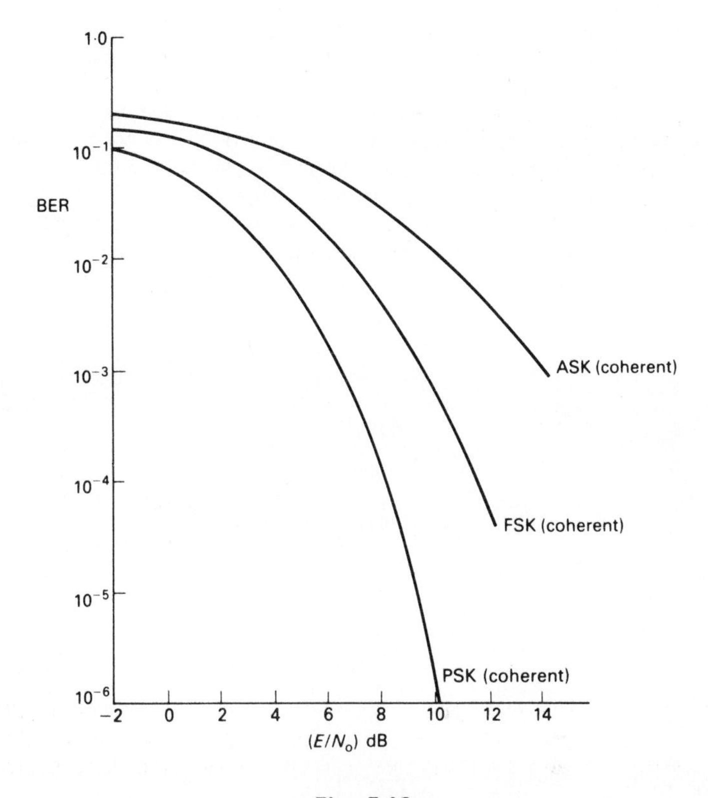

Fig. 5.18

5.12 Spread-spectrum modulation[29,30]

In a spread-spectrum system, the transmitted signal is spread over a wide frequency band which is much wider than the minimum bandwidth required to transmit the baseband information. The baseband signal of a few kilobauds,* e.g. 4800 bauds/s, is spread over a bandwidth which may be many MHz wide. This is achieved by using a wideband encoding signal which is modulated by the information to be transmitted.

For the transmission of binary information, the encoding signal is a digital code sequence whose bit rate is much higher than that of the binary information. Usually, a pseudo-noise code or PN code is used and it is simply

* The unit of signalling speed is known as the *baud*. See F. R. Connor, *Signals*, Edward Arnold (1982).

modulated by the binary information using modulo-2 addition as shown in Fig. 5.20. The modulated PN code sequence is then used to phase-shift a carrier wave using biphase PSK or quadriphase PSK.

The power spectrum of a purely random binary code sequence with a bit duration T_0 is derived in Appendix E. Since a finite PN code sequence must repeat itself in a regular fashion due to practical realisation by a generator, it has a line spectrum with the same envelope as that shown in Fig. 5.19. Typically, if the clock rate of the PN code sequence is 1 MHz, the main lobe bandwidth (null to null) is 2 MHz and a sidelobe bandwidth is 1 MHz wide.

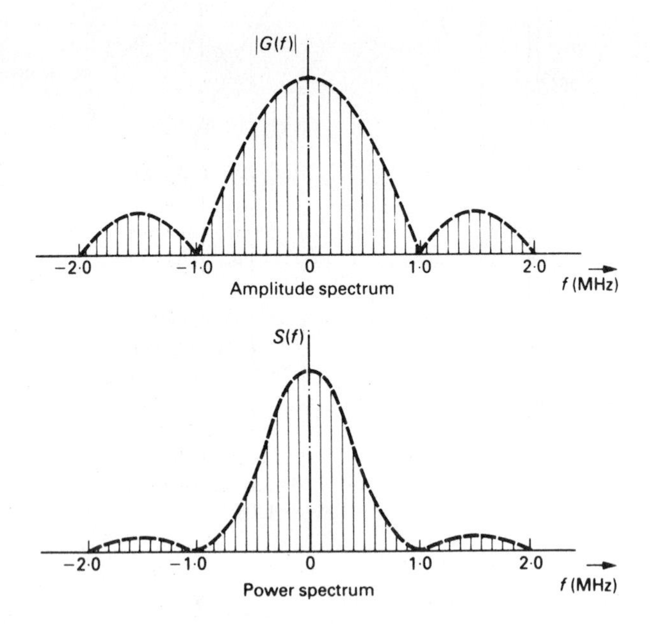

Fig. 5.19

The main application of spread-spectrum modulation is in the field of satellite communications. As satellite transmission power is very costly to provide, it must be used effectively and therefore digital information combined with some form of phase-shift-keying is widely used, especially in the field of military communications. Furthermore, by exchanging bandwidth for signal-to-noise ratio according to Shannon's information theory, the required communication capacity can still be achieved with a consequent saving in power.

In a military application, biphase balanced modulation may be employed as the carrier is suppressed and detection of it is therefore difficult. Moreover, more power is available for sending the useful coded information only. At the receiver, the signal is correlated with an exact replica of the PN code sequence used at the transmitter. The phase reversals of the original PN code are thus removed (being 360° for biphase), while the phase reversals due to the binary information alone are then extracted by a PRK demodulator. Hence, detection requires a correlation receiver of which a simple schematic diagram is given in Fig. 5.20.

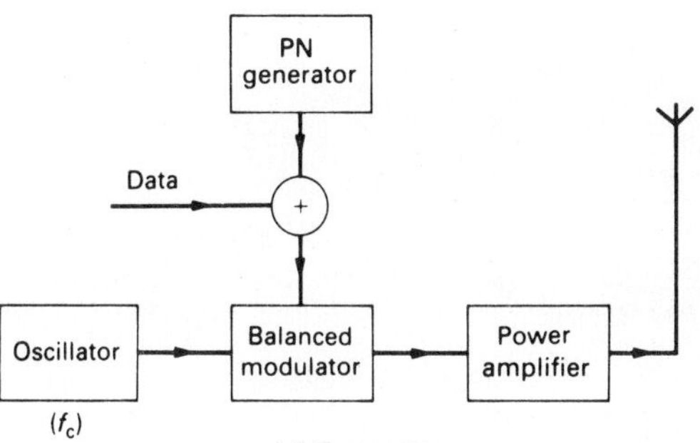

(a) Transmitter

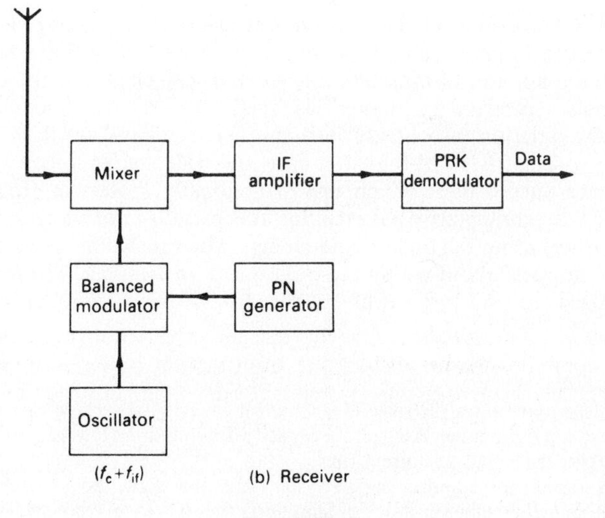

(b) Receiver

Fig. 5.20

The encoding signal for the data information is usually a maximal length pseudo-random code sequence derived from an n-stage shift register using feedback. The length of the codeword is $2^n - 1$ bits after which it repeats itself. The shift register is clocked by a frequency source typically at a rate of 1 Mbit/s. The output of the shift register is then combined with the data stream using modulo-2 addition by means of logic gates as shown in Fig. 5.21.

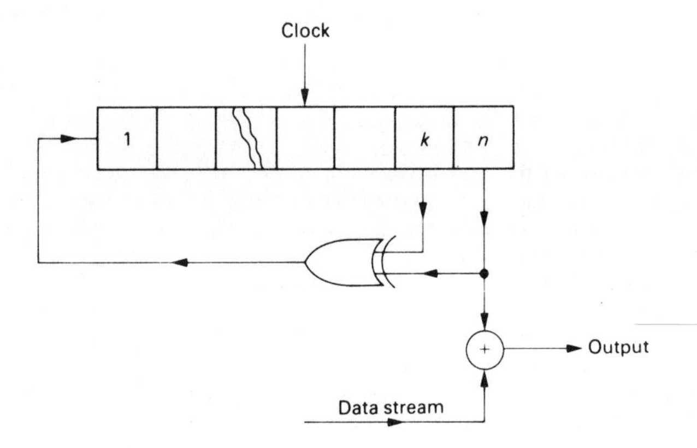

Fig. 5.21

In practice, there are a number of such code 'patterns' which may be generated for multichannel operation and the most suitable codes are those which yield the minimum amount of crosstalk due to cross-correlation at the receiver.

The spread-spectrum technique described above is also known as *direct sequence* modulation in contrast with another technique which is known as *frequency-hopping* (FH). In the latter case, the wide spectrum band chosen is divided into subchannels which are orthogonally spaced in frequency. A subchannel is momentarily selected for the pseudo-random technique (described earlier) using the digital data stream. The subchannel position is then moved or 'hopped' about within the wide spectrum available. The technique is mainly used in military applications for security and anti-jamming purposes.

Example 5.4

Shannon's theorem states the maximum channel capacity of a noisy communication channel. Explain the manner in which the capacity depends upon the channel bandwidth and also upon the signal-to-noise ratio.

A 'spread-spectrum' communication system uses low signal power and very large bandwidth. Explain the implications of Shannon's theorem in this case and suggest how such a system might be implemented. (C.E.I.)

Solution
The information capacity of a channel may be conveniently defined in terms of the maximum amount of information that can be transmitted correctly per unit time. Shannon in his famous theorem pointed out that if S is the average signal power and the noise in the channel is white Gaussian noise of mean power N in a bandwidth W, then, by suitable encoding, it is possible to transmit information as binary digits, with as small a frequency of errors as desired, at a rate given by

$$C = W \log_2 (1 + S/N) \text{ bits/s}$$

This important result for an ideal system implies that the value of C can only be increased by increasing W or S/N. Since an increase in S/N implies an increase in transmitter power, this in effect amounts to a power–bandwidth 'trade-off' in order to achieve any required value of C.

However, because of the logarithmic relationship involved, the trade-off is not a simple one and, in general, is unfavourable both in the ideal case and in most practical systems. For example, a reduction in bandwidth by five-fold in the ideal case would require an increase in transmitter power of about sixty-four times and, for many practical systems, such a power increase would be prohibitive. Nevertheless, the significance of this exchange is well known in broadband systems, such as FM and PCM, which are capable of giving a considerable improvement in S/N performance at the expense of a wide bandwidth, though they still fall short of the ideal system.

In a spread-spectrum communication system, it is possible to operate with a low S/N ratio and a very large bandwidth by using, as a basis, Shannon's previous result. Since

$$C = W \log_2 (1 + S/N)$$

we have

$$C/W = \ln (1 + S/N)/\ln 2$$

or

$$C/W \simeq 1{\cdot}44\, S/N$$

Hence

$$W \simeq \frac{CN}{1{\cdot}44\, S}$$

Assuming a data rate of $C = 3$ kbits/s and a value of $S/N = 0{\cdot}01$, this yields

$$W \simeq \frac{3 \times 10^3 \times 100}{1{\cdot}44} \simeq 0{\cdot}2 \text{ MHz}$$

and so operation at the very low S/N ratio of -20 dB is possible by spreading the signal power over the much wider bandwidth of about 0·2 MHz.

Hence, this is a trade-off involving a considerable reduction in S/N ratio by the use of a very much wider bandwidth than the original data information, with a consequent saving in transmitter power.

A practical application of this technique is in transmitting data from a communications satellite. Since transmitter power in the satellite is expensive, it must be used very economically to achieve the optimum performance. An implementation of this system was described in Section 5.12 and a schematic diagram is given in Fig. 5.20.

6

Demodulation

The process of extracting or recovering the information from a modulated signal is known as demodulation or detection. As modulated signals are either of the analogue or digital type, the demodulation techniques used are different in each case. In addition, demodulation of AM and FM signals require different methods because of their different characteristics. However, in general, linear demodulation is essential in both AM and FM receivers in order to minimise signal distortion. A brief description of AM and FM receivers is given in Appendix F.

6.1 AM detectors[31,32]

The demodulation of an AM signal requires a non-linear or linear detector. A typical non-linear detector is the square-law detector, while the two main types of linear detectors are the non-coherent or *envelope* detector and the coherent or *synchronous* detector.

The envelope detector, because of its simplicity, is by far the most common type used for AM signals. This applies equally well to speech, music, or video signals where, in the case of video signals only, VSB operation is employed.

Coherent or synchronous detectors may also be used for AM signals, but more usually they are employed for DSBSC or SSBSC signals. The synchronous detector is by far the more critical in its operation than is the envelope detector, since it depends upon exact carrier synchronisation for its proper operation.

Square-law detector

A semiconductor diode in series with a resistive load as shown in Fig. 6.1(a) has a non-linear input–output characteristic in the forward direction. This is illustrated in Fig. 6.1(b) and it is expressed by the equation

$$i = av + bv^2$$

where i is the output current, v is the input voltage, and a and b are arbitrary constants.

For small input voltages, the device can be used to detect weak signals as in radar or for the demodulation of AM signals. In the latter case, the input

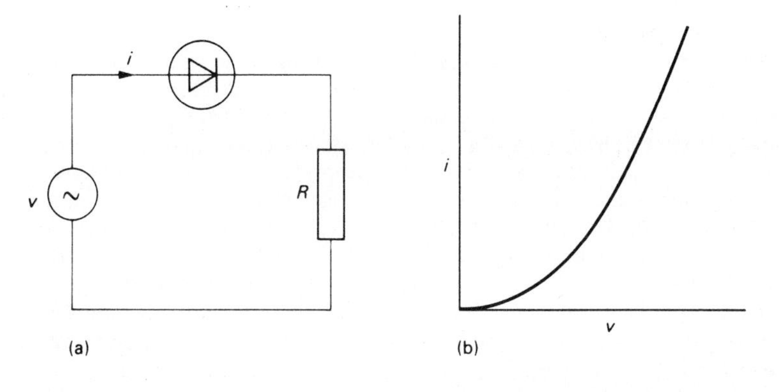

Fig. 6.1

voltage for a single modulating tone is given by

$$v = V_c(1 + m \sin \omega_m t) \sin \omega_c t$$

where V_c is the peak carrier voltage, m is the depth of modulation, ω_m is the angular modulating frequency, and ω_c is the angular carrier frequency. Substituting for v in the previous equation yields

$$i = aV_c(1 + m \sin \omega_m t) \sin \omega_c t + bV_c^2(1 + m \sin \omega_m t) \sin^2 \omega_c t$$

The output current at the fundamental modulating frequency arises from the second term and is given by

$$i_1 = mbV_c^2 \sin \omega_m t$$

while the second harmonic term has a magnitude of

$$i_2 = \frac{m^2 bV_c^2}{4} \cos 2\omega_m t$$

From these we obtain the important peak ratio

$$\frac{\text{second harmonic term}}{\text{fundamental term}} = \frac{m^2 bV_c^2/4}{mbV_c^2} = \frac{m}{4}$$

and so the depth of modulation must be small to minimise the harmonic distortion.

An important application of this detector is in radar systems where waveform distortion is of no consequence. The main concern is to *detect* a weak pulse signal reflected from a distant target and pulse integration is then used to improve the signal-to-noise ratio.

Envelope detector

Since the shape of the envelope of an AM signal is similar to that of the modulating signal, a circuit which can follow the envelope waveform is the linear diode detector. Though the diode is a non-linear device, its voltage–current characteristic may be regarded as essentially linear for *large* input signals. The output signal is therefore proportional to the input signal.

The basic circuit shown in Fig. 6.2 consists of a diode (crystal or valve) in series with an *RC* load. As such it behaves as a half-wave rectifier and the capacitor *C* is initially charged and then discharged with a time-constant *RC*. If the time-constant is correct, the output voltage will follow the modulation envelope and consists of the audio signal together with a d.c. voltage. The latter is removed by the blocking capacitor C_1 and the audio signal appears across the resistor R_1.

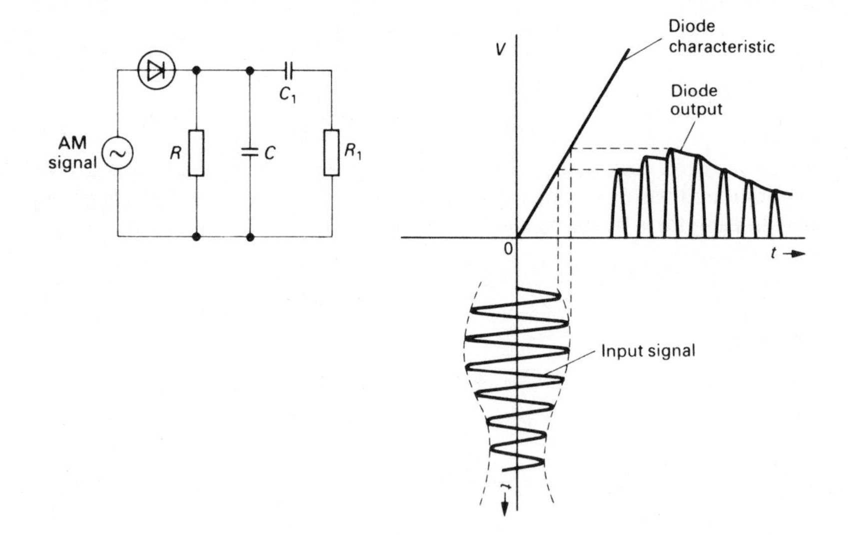

Fig. 6.2

To obtain the correct value of *RC*, let the voltage on the capacitor at time *t* be $v = V_0 e^{-t/RC}$, where V_0 is the initial value at the *peak* of the envelope. Differentiating with respect to time yields

$$-\frac{dv}{dt} = \frac{V_0 e^{-t/RC}}{RC} = \frac{v}{RC}$$

Since the decrease in capacitor voltage must follow the modulation envelope, which is of the form $v = kV_c(1 + m\sin\omega_m t)$ for a single modulating frequency f_m and in which k is the diode efficiency, then for no distortion we must have

$$-\frac{dv}{dt} = \frac{d}{dt}[kV_c(1 + m\sin\omega_m t)]$$

or

$$-\frac{dv}{dt} = -km\omega_m V_c \cos\omega_m t$$

To ensure that the rate of discharge of the capacitor voltage, i.e. $-dv/dt$, is sufficiently great, it is more practical to use the inequality

$$-\frac{dv}{dt} \geqslant -km\omega_m \cos\omega_m t$$

or

$$v/RC \geqslant -km\omega_m \cos\omega_m t$$

with

$$\frac{kV_c}{RC}(1 + m\sin\omega_m t) \geqslant -km\omega_m \cos\omega_m t$$

or

$$RC \leqslant \frac{1 + m\sin\omega_m t}{-m\omega_m \cos\omega_m t}$$

By differentiating the right hand side of the last expression and equating it to zero, the expression for RC is shown to be a minimum when $\sin\omega_m t = -m$ and $\cos\omega_m t = -\sqrt{1 - m^2}$. Hence

$$RC \leqslant \frac{\sqrt{1 - m^2}}{m\omega_m}$$

In practice, this condition can be met for low values of m. Usually, $m = 0{\cdot}3$ to 0·4 for radio broadcast systems transmitting music to ensure a minimum amount of envelope distortion on demodulation.

A more detailed analysis of the circuit for a modulated carrier in the presence of noise is given in another volume* of this series.

Example 6.1

Discuss the merits in radio receivers of an envelope detector which uses a diode with an RC load.

The input to an envelope detector is an amplitude-modulated wave from which one side-frequency has been suppressed. If the original modulation depth m is low, show that the second harmonic distortion of the output from the detector (assumed to be linear) is approximately 12·5m per cent.

* See F. R. Connor, *Noise*, Edward Arnold (1982).

Explain how distortion from this cause is reduced in single-sideband transmissions. (U.L.)

Solution

The merits of an envelope detector in a radio receiver are:

1. It simplifies receiver construction, since no sophisticated or critical circuits need to be adjusted.
2. It reduces receiver costs and so enables the widespread use of such equipment.
3. It operates as a linear device for large input signals with little distortion and the output is sufficient to drive an amplifier.

Problem

The input to the detector $v_i(t)$ is given by

$$v_i(t) = V_c \sin \omega_c t + \frac{mV_c}{2} \cos(\omega_c - \omega_m)t \quad \text{(LSF only)}$$

$$= V_c \sin \omega_c t + \frac{mV_c}{2}\{\cos \omega_c t \cos \omega_m t + \sin \omega_c t \sin \omega_m t\}$$

$$= V_c \sin \omega_c t\left[1 + \left(\frac{m}{2}\right)\sin \omega_m t\right] + \frac{mV_c}{2} \cos \omega_c t \cos \omega_m t$$

or

$$v_i(t) = A \sin \omega_c t + B \cos \omega_c t = \sqrt{A^2 + B^2} \sin(\omega_c t + \phi)$$

where $A = V_c[1 + (m/2)\sin \omega_m t]$, $B = (mV_c/2)\cos \omega_m t$, and $\phi = \tan^{-1}(B/A)$.

The output of the detector corresponds to the *envelope* of $v_i(t)$ and is given by

$$v_o(t) = \sqrt{A^2 + B^2} = \sqrt{V_c^2(1 + m \sin \omega_m t + m^2/4)}$$

$$= V_c[1 + m^2/4 + m \sin \omega_m t]^{1/2}$$

$$= V_c\sqrt{1 + m^2/4}\left[1 + \frac{m}{(1 + m^2/4)} \sin \omega_m t\right]^{1/2}$$

$$= V_c\sqrt{1 + m^2/4}\left[1 + \frac{m}{2(1 + m^2/4)} \sin \omega_m t - \frac{m^2}{16(1 + m^2/4)^2} + \frac{m^2}{16(1 + m^2/4)^2} \cos 2\omega_m t + \ldots\right]$$

if $m \ll 1$.

The second harmonic distortion is given by the fourth term and its amplitude is $m^2/16(1 + m^2/4)^2 \simeq m^2/16$ if m is small. The amplitude of the fundamental component in the same expression is $\simeq m/2$, hence the second harmonic distortion is given by

$$\text{percentage distortion} = \frac{m^2}{16(m/2)} \times 100 = 12{\cdot}5m\,\%$$

Distortion from this cause may be reduced in an SSB transmission by adding an additional carrier signal at the receiver so that the ratio of the sideband voltage to the total carrier voltage *before* demodulation is small.

Example 6.2

Determine the filtered output from a quadratic detector ($v_o = Av_i^2$) when the input is

(a) the upper side-frequency component only of the amplitude-modulated wave $e = \hat{E}(1 + m\cos pt)\sin\omega t$ together with the voltage $\hat{E}_c \sin(\omega t + \phi)$ from a local oscillator,

(b) both side-frequency components of the same modulated wave together with the voltage from the local oscillator.

Comment on the harmonic distortion of the output signal from the detector and the significance of the angle ϕ. (U.L.)

Solution

(a) The frequency components of the AM wave are given by

$$e = \hat{E}(1 + m\cos pt)\sin\omega t = \hat{E}\sin\omega t + m\hat{E}\sin\omega t\cos pt$$

or

$$e = \hat{E}\sin\omega t + \frac{m\hat{E}}{2}\{\sin(\omega + p)t + \sin(\omega - p)t\}$$

The upper side-frequency component is therefore

$$e_{usf} = \frac{m\hat{E}}{2}\sin(\omega + p)t$$

The input to the quadratic detector is

$$v_i = \frac{m\hat{E}}{2}\sin(\omega + p)t + \hat{E}_c\sin(\omega t + \phi)$$

and the output from the detector is

$$v_o = A\left[\frac{m\hat{E}}{2}\sin(\omega + p)t + \hat{E}_c\sin(\omega t + \phi)\right]^2$$

$$= A\left[\frac{m^2\hat{E}^2}{4}\sin^2(\omega + p)t + m\hat{E}\hat{E}_c\sin(\omega + p)t\sin(\omega t + \phi) + \hat{E}_c^2\sin^2(\omega t + \phi)\right]$$

or

$$v_o = A\left[\frac{m^2\hat{E}^2}{8} - \frac{m^2\hat{E}^2}{8}\cos 2(\omega + p)t + \frac{m\hat{E}\hat{E}_c}{2}\{\cos(pt - \phi) - \cos(2\omega t + pt + \phi)\}\right.$$
$$\left. + \frac{\hat{E}_c^2}{2} - \frac{\hat{E}_c^2}{2}\cos 2(\omega t + \phi)\right]$$

Comments

1. The output contains d.c. terms, a modulation term, and second harmonics of the input frequencies.
2. There is no harmonic distortion of the modulation.
3. The modulation term is simply delayed by the phase angle ϕ and synchronisation of the local oscillator is not essential.

(b) The input to the quadratic detector is

$$v_i = \hat{E}_c \sin(\omega t + \phi) + m\hat{E} \sin \omega t \cos pt$$

and the output from the detector is

$$\begin{aligned} v_o &= A[\hat{E}_c^2 \sin^2(\omega t + \phi) + 2m\hat{E}_c\hat{E} \sin(\omega t + \phi) \sin \omega t \cos pt \\ &\quad + m^2\hat{E}^2 \sin^2 \omega t \cos^2 pt] \\ &= A\left[\frac{\hat{E}_c^2}{2} - \frac{\hat{E}_c^2}{2} \cos 2(\omega t + \phi)\right. \\ &\quad + 2m\hat{E}_c\hat{E}\{(\sin \omega t \cos \phi + \cos \omega t \sin \phi) \sin \omega t \cos pt\} \\ &\quad \left. + \frac{m^2\hat{E}^2}{4}\{(1 - \cos 2\omega t)(1 + \cos 2pt)\}\right] \end{aligned}$$

Hence

$$\begin{aligned} v_o = A&\left[\frac{\hat{E}_c^2}{2} - \frac{\hat{E}_c^2}{2} \cos 2(\omega t + \phi)\right. \\ &+ m\hat{E}_c\hat{E}\{(1 - \cos 2\omega t) \cos pt \cos \phi + \sin 2\omega t \cos pt \sin \phi\} \\ &\left. + \frac{m^2\hat{E}^2}{4}\{1 - \cos 2\omega t + \cos 2pt - \cos 2\omega t \cos 2pt\}\right] \end{aligned}$$

or

$$\begin{aligned} v_o = A&\left[\frac{\hat{E}_c^2}{2} - \frac{\hat{E}_c^2}{2} \cos 2(\omega t + \phi) + m\hat{E}_c\hat{E} \cos pt \cos \phi\right. \\ &- \frac{m\hat{E}_c\hat{E}}{2} \cos \phi\{\cos(2\omega + p)t + \cos(2\omega - p)t\} \\ &+ \frac{m\hat{E}_c\hat{E}}{2} \sin \phi\{\sin(2\omega + p)t - \sin(2\omega - p)t\} \\ &+ \frac{m^2\hat{E}^2}{4} - \frac{m^2\hat{E}^2}{4} \cos 2\omega t + \frac{m^2\hat{E}^2}{4} \cos 2pt \\ &\left. - \frac{m^2\hat{E}^2}{8}\{\cos(2\omega + 2p)t + \cos(2\omega - 2p)t\}\right] \end{aligned}$$

Comments

1. The output contains d.c. terms, a modulation term, and harmonics of the input frequencies.
2. The modulation term $Am\hat{E}_c\hat{E} \cos pt \cos \phi$ is a maximum when $\phi = 0$ and vanishes when $\phi = \pi/2$. Hence, the local oscillator must be stable and exactly synchronised.
3. There is some second harmonic distortion in the modulation. Its value relative to the fundamental is

$$\frac{Am^2\hat{E}^2}{4} \bigg/ Am\hat{E}_c\hat{E} = \frac{m\hat{E}}{4\hat{E}_c}$$

VSB demodulation

A VSB television signal is usually demodulated by an envelope detector for both simplicity and economy. One way of producing a VSB signal is by adding a smaller pair of sidebands to an AM signal but in phase *quadrature* with the carrier. This is illustrated in Fig. 6.3, where the AM signal is shown in Fig. 6.3(a), the same signal with the quadrature sidebands in Fig. 6.3(b), and the resultant VSB signal in Fig. 6.3(c).

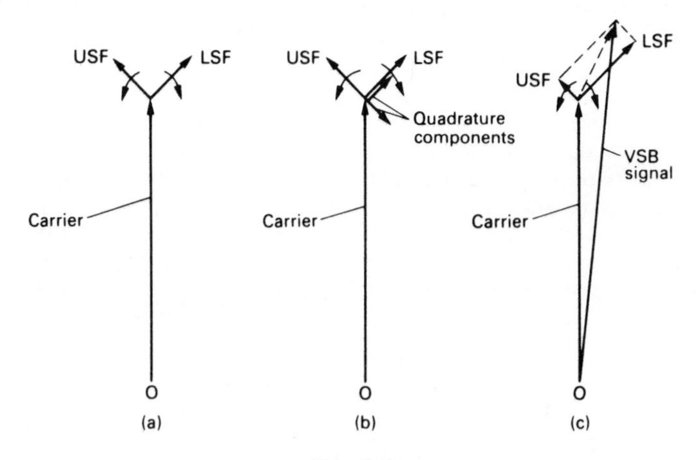

Fig. 6.3

The expression for an AM signal was obtained in Chapter 2 as

$$v(t) = V_c(1 + m \sin \omega_m t) \sin \omega_c t$$

where the term $mV_c \sin \omega_m t \sin \omega_c t$ represents the pair of *in-phase* sidebands. The smaller quadrature pair of sidebands is given by $kmV_c \cos \omega_m t \cos \omega_c t$, where $0 < k < 1$, and the input to the detector is a VSB signal given by

$$v_i(t) = V_c[(1 + m \sin \omega_m t) \sin \omega_c t + km \cos \omega_m t \cos \omega_c t]$$

or
$$v_i(t) = V_c\{[1 + p(t)] \sin \omega_c t + q(t) \cos \omega_c t\}$$

where $p(t) = m \sin \omega_m t$ and $q(t) = km \cos \omega_m t$.

Hence
$$v_i(t) = V_c\left[\sqrt{A^2 + B^2} \sin(\omega_c t + \phi)\right]$$

where $A = 1 + p(t)$, $B = q(t)$, and $\phi = \tan^{-1}(B/A)$.

The input to the detector is an angle-modulated wave whose amplitude also changes. The detector responds only to the amplitude changes and so the output from an ideal detector is given by

$$v_o(t) = V_c\left[\sqrt{A^2 + B^2}\right] = AV_c\sqrt{1 + (B/A)^2}$$

so $$v_o(t) \simeq AV_c \qquad \text{if } B \ll A$$

i.e. $$v_o(t) \simeq V_c(1 + m \sin \omega_m t)$$

since $A = (1 + m \sin \omega_m t)$.

This result is similar to the demodulated envelope of an AM signal and there is no distortion if $B \ll A$, i.e. the quadrature component $q(t)$ must be small. Since $q(t)$ depends upon km, m can be increased in practice to increase sideband power (or reduce carrier power) provided the product km is still small.

In addition, a diode detector can also be used for demodulation.

Synchronous detector

The action of this detector is somewhat similar to that of a multiplicative mixer whereby two signals are multiplied by a non-linear device, to give sum and difference frequencies in the output. Hence, in the synchronous detector the incoming signal, which is generally a DSBSC or SSBSC signal, is multiplied with a local carrier signal in a non-linear circuit. The output is then passed through an appropriate low-pass filter to yield the modulation required. Fig. 6.4 shows a schematic circuit for an incoming signal $v_i = f(t)$ and a local oscillator signal $v_2 = V_2 \sin \omega_c t = \sin \omega_c t$ if $V_2 = 1$ volt, for convenience.

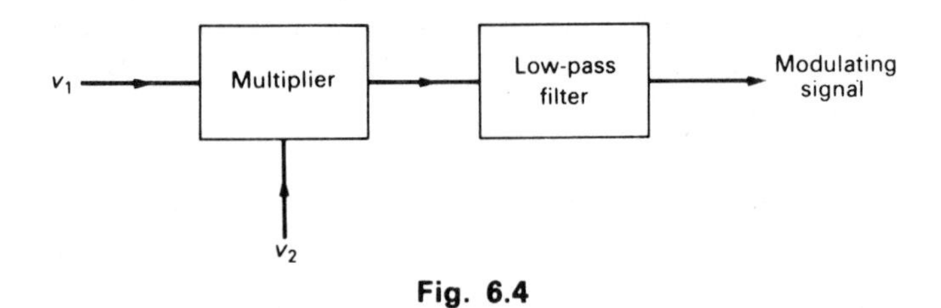

Fig. 6.4

For the case of a DSBSC signal we have

$$v_1 = f(t) = mV_c \sin \omega_m t \sin \omega_c t$$
$$v_2 = \sin \omega_c t$$

where $m = V_m/V_c$ is the depth of modulation. Since a multiplier performs the operation $v_o = kv_1v_2$, where k has the dimension volt^{-1}, we obtain

$$v_o = kv_1v_2 = kmV_c \sin \omega_m t \sin^2 \omega_c t = kV_m \sin \omega_m t \sin^2 \omega_c t$$

$$= \frac{kV_m \sin \omega_m t}{2}[1 - \cos 2\omega_c t]$$

or $$v_o = \left(\frac{kV_m}{2}\right) \sin \omega_m t - \left(\frac{kV_m}{2}\right) \sin \omega_m t \cos 2\omega_c t$$

The low-pass filter subsequently separates out the modulation term

$$\left(\frac{kV_m}{2}\right)\sin\omega_m t$$

For the case of an SSBSC signal we have

$$v_1 = f(t) = \left(\frac{kmV_c}{2}\right)\cos(\omega_c - \omega_m)t$$

for the lower side-frequency only and $v_2 = \sin\omega_c t$. Hence

$$v_o = kv_1v_2 = \left(\frac{kmV_c}{2}\right)\cos(\omega_c - \omega_m)t\sin\omega_c t$$

or

$$v_o = \frac{kmV_c}{4}[\sin\omega_m t + \sin(2\omega_c - \omega_m)t]$$

and the modulation term is $(kmV_c/4)\sin\omega_m t$ which can be separated out by a low-pass filter.

However, for synchronous or coherent detection to be effective, synchronisation of the locally injected carrier must be exact. The effect of a phase or frequency shift can be considered by assuming that the injected local carrier signal is $v_2 = \sin(\omega_c' t + \phi)$, where $\omega_c' = (\omega_c + \delta\omega_c)$. For both DSBSC and SSBSC signals, it is shown in Appendix G that distortion or phase delay can occur and, in particular, when $\phi = \pi/2$, the modulation disappears for a DSBSC signal.

Hence, to ensure exact synchronisation, a pilot carrier about 26 dB below the normal value is transmitted. At the receiver, the pilot carrier is used to lock the local oscillator or, alternatively, it may be amplified and used as the local carrier signal.

Example 6.3

One input to the multiplier shown in Fig. 6.5 is a carrier wave of rms amplitude $\sqrt{2}$ volt, the other input is the same carrier after transmission through a modulator to which a low-frequency, periodic, two-level [0, 1] signal is applied. When the signal level is 0, the modulator transmits the carrier unchanged in any way, so that the second input to the multiplier is then $\sqrt{2}$ volt. The low-pass filter suppresses all components at and above the carrier frequency and transmits all lower frequencies without loss.

Deduce for each of the following cases the waveform and its amplitude at the output of the filter

(a) the signal phase modulates the carrier with a phase deviation of $+\pi/2$ radians when the signal level is 1,

(b) the signal frequency modulates the carrier with a frequency deviation of ± 120 Hz when the signal level is 1,

(c) the signal amplitude modulates the carrier so that the output from the modulator is zero when the signal level is 1. (U.L.)

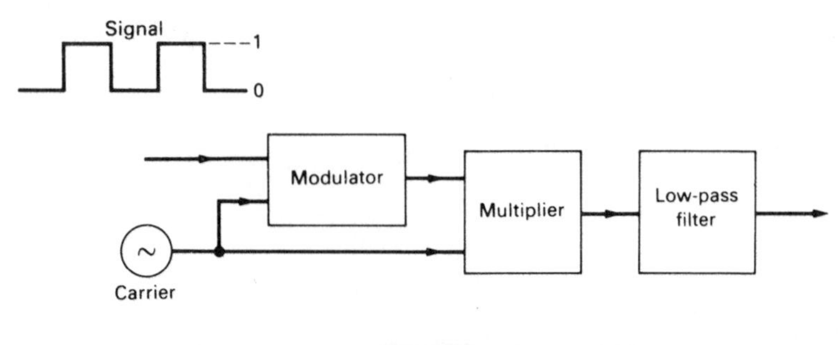

Fig. 6.5

Solution
Let the output signal from the modulator be s_1 and let that from the carrier oscillator be $s_2 = 2\sin\omega_c t$, where $\omega_c/2\pi$ is the carrier frequency. The output of the multiplier is $s_1 \times s_2$ and, after filtering, let the output from the low-pass filter be s_o. Hence, we have

(a) $s_1 = 2\sin(\omega_c t + \phi)$ where $\phi = 0$ or $\pi/2$ (alternate intervals)
$s_2 = 2\sin\omega_c t$
$s_1 \times s_2 = 4\sin(\omega_c t + \phi)\sin\omega_c t = 2\{\cos\phi - \cos(2\omega_c t + \phi)\}$
$s_o = 2\cos\phi = 0$ or 2 V pulse (alternate intervals)

(b) $s_1 = 2\sin 2\pi f_i t$ where $f_i = (f_c + 120)$ Hz or f_c (alternate intervals)
$s_2 = 2\sin 2\pi f_c t$
$s_1 \times s_2 = 4\sin 2\pi f_i t \sin 2\pi f_c t = 2\{\cos 2\pi(f_i - f_c)t - \cos 2\pi(f_i + f_c)t\}$
$s_o = 2\cos 2\pi(f_i - f_c)t$ = 120 Hz cosine wave or 2V d.c. (alternate intervals)

(c) $s_1 = V_c \sin\omega_c t$ where $V_c = 0$ or 2V (alternate intervals)
$s_2 = 2\sin\omega_c t$
$s_1 \times s_2 = 2V_c \sin^2\omega_c t = V_c(1 - \cos 2\omega_c t)$
$s_o = V_c = 0$ or 2 V pulse (alternate intervals)

Detector performance
The detection of signals in a background of noise is of primary importance in communication systems. Hence, the performance of a detector is usually described in terms of its input signal-to-noise ratio (S_i/N_i) and its output signal-to-noise ratio (S_o/N_o), both of which are expressed in decibels.

In Appendix H the analyses for the square-law, linear, and synchronous detectors are described, assuming narrowband Gaussian noise and a sinusoidal carrier as the input signal. The results are shown plotted in Fig. 6.6. For low (S_i/N_i) ratios, i.e. $(S_i/N_i) \simeq 1$, the square-law and linear detectors have a similar performance, but for large (S_i/N_i) ratios the linear detector has a 3 dB advantage over the square-law detector. However, for $(S_i/N_i) < 1$ the output

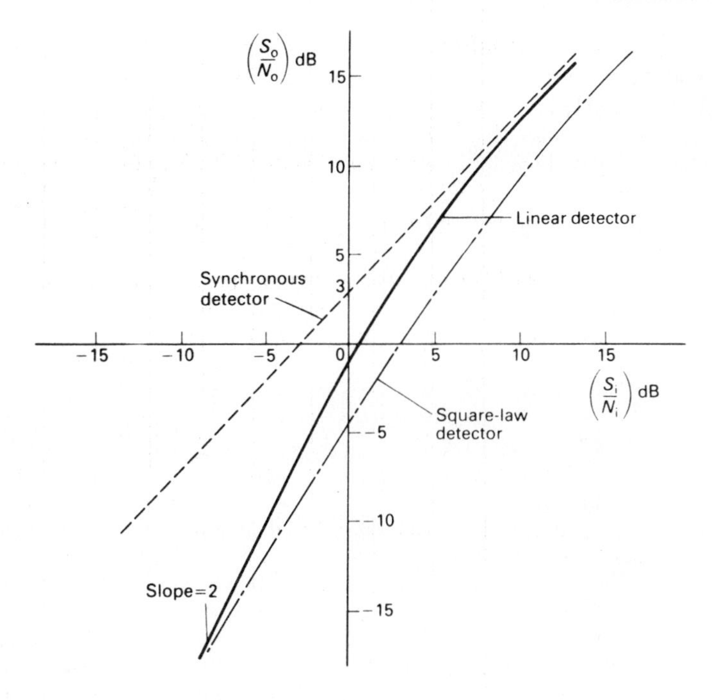

Fig. 6.6

signal-to-noise ratio is so poor that the signal is lost in a noisy background which is known as the AM *threshold effect*.

The best performance is achieved by the synchronous detector even down to low (S_i/N_i) ratios. For ratios of $(S_i/N_i) < 1$, it has a 3 dB advantage over the other detectors and there is no threshold effect. At large signal-to-noise ratios, the linear and synchronous detectors perform equally well.

6.2 FM discriminators

In order to extract the modulation from an FM wave, changes in frequency must be converted to corresponding amplitude changes. Circuits used for this purpose are usually called discriminators and the two most widely used types are the Foster–Seeley circuit and the ratio detector. The former generally gives better linearity but must be preceded by a limiter, while the latter combines both limiting and discrimination in the same circuit. It is largely used in domestic receivers.

Foster–Seeley discriminator[33]
The typical circuit shown in Fig. 6.7(a) consists of two diode detectors placed back to back and fed from a common secondary winding. The primary voltage V_P is fed via a capacitor to the centre tap of the secondary winding and so the secondary voltages V_1 and V_2 are in antiphase.

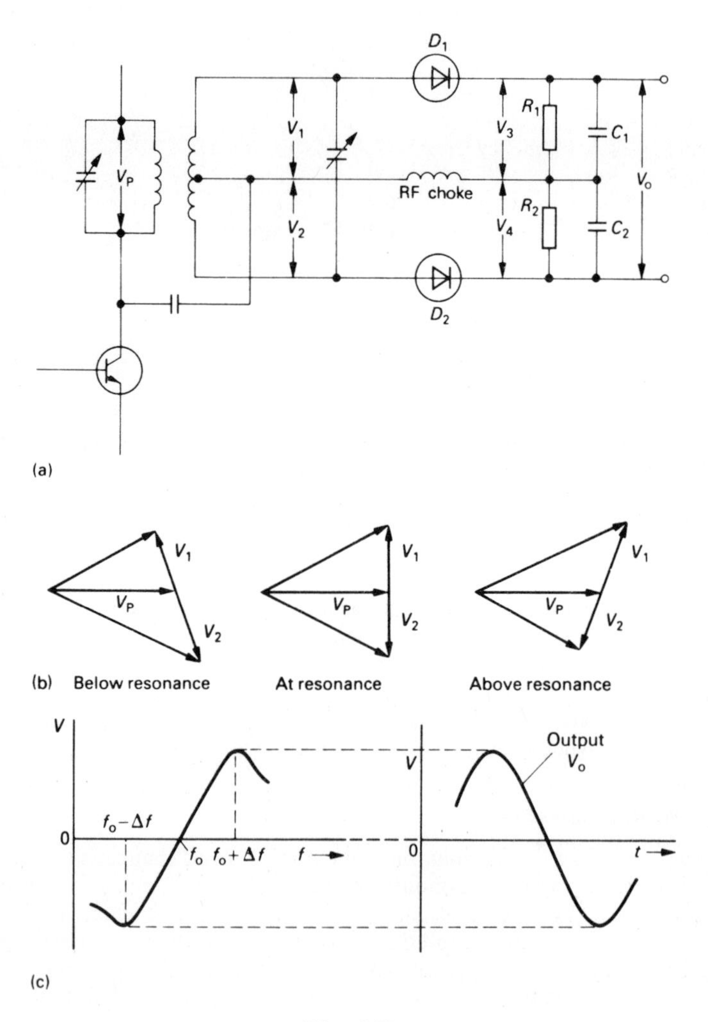

Fig. 6.7

The primary current I_P is in phase with V_P at resonance and induces voltages V_1 and V_2 in the secondary circuit, which are in quadrature with V_P since $V_1 = -j\omega MI_P = \omega MI_P\angle -90°$, where M is the mutual inductance between the windings.

From the phasor diagrams shown in Fig. 6.7(b) we see that voltages $(V_P + V_1)$ and $(V_P + V_2)$ are applied to the two diodes D_1 and D_2 respectively. The d.c. voltages V_3 and V_4 are equal and opposite at resonance and so V_o is zero. Above resonance, the primary circuit is inductive and so I_P lags V_P by a small angle. Voltage V_3 is greater than V_4 and V_o is positive. Below resonance, the primary circuit is capacitive and I_P leads V_P by a small angle. Voltage V_4 is then greater than V_3 and V_o is negative.

Hence, as the frequency changes, the output characteristic is the familiar S-shaped curve shown in Fig. 6.7(c) which is linear over a range of ± 75 kHz. Since the input frequency varies according to the modulation, the output voltage is the modulating signal with little distortion.

Ratio detector[34]

The circuit shown in Fig. 6.8(a) is in some respects similar to Fig. 6.7(a) but with diode D_2 reversed and an additional large capacitor C_0 placed across the resistors R_3 and R_4. Moreover, a tertiary winding is used instead of applying the primary voltage to the centre tap of the secondary winding.

In the phasor diagram shown in Fig. 6.8(b), V_T is the voltage in the tertiary winding. At resonance, the voltages $(V_T + V_1)$ and $(V_T + V_2)$, which are equal, are applied to the diodes *in series*. This causes a current to flow in R_3 and R_4 and a voltage V appears across them. Usually, $R_3 = R_4$ and so the output voltage $V_o = (V/2 - V_3) = (V_4 - V/2)$. Hence

$$2V_o = (V/2 - V_3) + (V_4 - V/2) = V_4 - V_3$$

or

$$V_o = (V_4 - V_3)/2$$

Also, if $C_3 = C_4$, then $V_3 = V_4$ at resonance and $V_o = 0$. Above resonance, $V_3 > V_4$ and V_o is negative while, below resonance, $V_3 < V_4$ and V_o is positive. The output voltage V_o therefore varies in amplitude as the input signal frequency changes according to the modulation. Since the ratio of voltages V_3/V_4 varies with the modulation, hence the name ratio detector.

An additional feature of the circuit is its limiting action. By connecting a large capacitor C_0 across R_3 and R_4, amplitude changes at the secondary winding cause C_0 to charge or discharge. However, as the time-constant $(R_3 + R_4)C_0$ is about 0·1 s, rapid amplitude fluctuations due to noise or signal fading are 'smoothed out' and the voltage across C_0 is fairly constant. The capacitor therefore acts as a large storage tank, the 'level' of which remains fairly constant. Hence, the output signal V_o is not affected by amplitude changes but only by frequency changes due to the modulating signal.

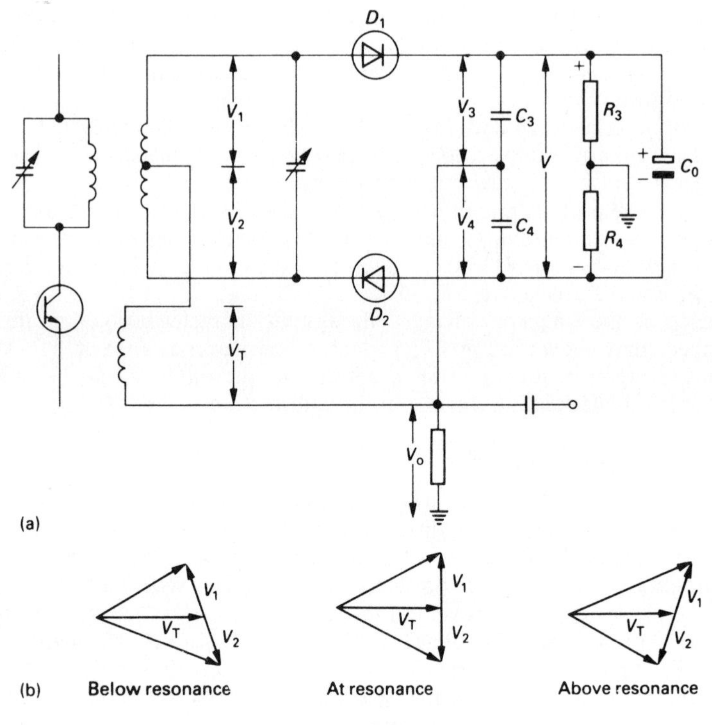

Fig. 6.8

6.3 Phase demodulation

A phase-modulated wave may be demodulated by means of an FM receiver followed by an integrating network. Since the FM receiver gives an output voltage proportional to the frequency deviation Δf, where $\Delta f = f_m \Delta\phi$ and f_m is the modulating frequency, the output must be passed through a $1/f_m$ network to obtain a voltage proportional to the phase deviation $\Delta\phi$ which corresponds to the modulating signal. The technique is illustrated in Fig. 6.9.

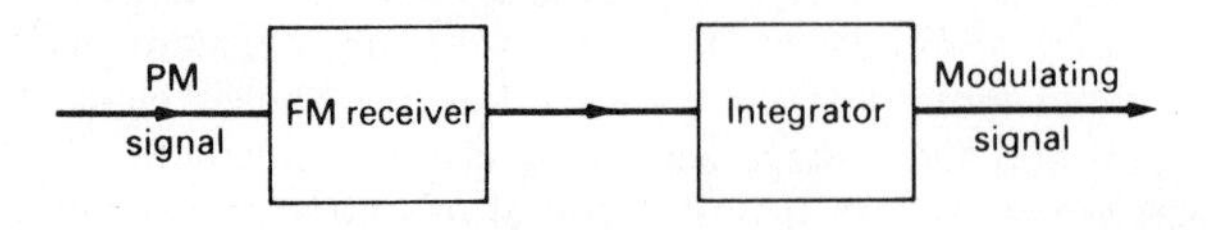

Fig. 6.9

6.4 Phase-sensitive detector[35]

The circuit shown in Fig. 6.10 may be used to obtain automatic frequency control of an oscillator because it is sensitive to phase changes caused by the drift in frequency of the oscillator. This is achieved by comparing the phase of the drifting oscillator with that of a very stable local oscillator. As the frequency of the input signal drifts, its phase changes continuously relative to the stable oscillator and so the output voltage from the detector will change in much the same way as for the Foster–Seeley discriminator. This output voltage is then applied via a feedback loop to control the frequency of the drifting oscillator. Alternatively, the circuit can be used as part of a phase-locked loop FM demodulator.

A similar circuit is extensively used in control systems, where the input corresponds to an error signal from a servo-system and the output is used via a feedback servo-loop to reduce the error signal.

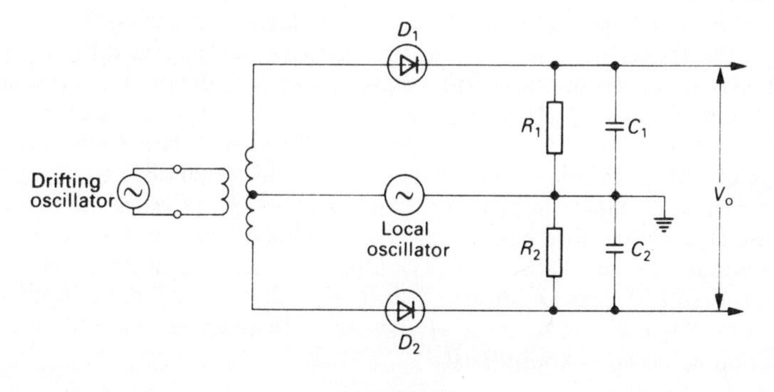

Fig. 6.10

6.5 Feedback demodulators[36]

The FM threshold[37] can be reduced by using a feedback demodulator which employs a negative feedback loop, such as the frequency-locked loop (FLL) or the phase-locked loop (PLL).

Frequency-locked loop (FLL)

This loop functions essentially as a narrowband tracking filter which follows the input frequency variations very closely. A typical circuit arrangement is shown in Fig. 6.11.

The input signal is mixed with the output from a variable control oscillator (VCO) and the difference frequency $(\omega_i - \omega_o) = \omega_c$ is centred at the middle of the band-pass filter (at frequency lock) before being differentiated by the

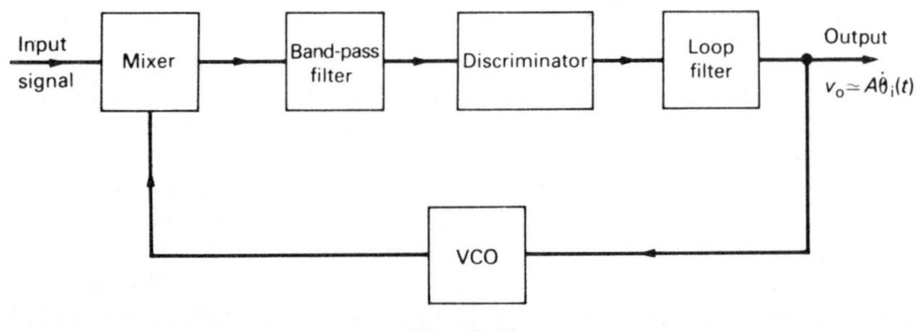

Fig. 6.11

discriminator. The output from the discriminator consists of various low-frequency components and is passed through a narrowband, low-pass loop filter to control the frequency of the VCO, such that $(\omega_i - \omega_o) = \omega_c$ at frequency lock.

It is shown in Appendix I that the control voltage from the loop filter is given by $v_o \simeq A\dot{\theta}_i(t)$, where $A = 1/\beta$, β is the feedback factor, and $\dot{\theta}_i(t)$ is the first differential of the instantaneous phase angle. Hence, if the input is a frequency-modulated signal and the loop filter has a bandwidth equal to twice the baseband modulating signal, its output follows the modulation and the circuit acts as an FM demodulator. However, as shown in Appendix I, the bandwidth of the band-pass filter can be considerably smaller than that of the FM signal and the circuit therefore behaves as a narrowband tracking filter.

Consequently, its output (S/N) ratio is much larger than that of a conventional FM demodulator and so it can still demodulate an input signal with a (S/N) ratio below the FM threshold. In practice, the FM threshold reduction achieved is about 6 dB.

Phase-locked loop (PLL)[38]

Unlike the FLL which can achieve a frequency lock only, this loop is capable of achieving both a frequency lock and a phase lock. It is used extensively in many communication circuits and a typical arrangement is shown in Fig. 6.12.

The PLL circuit is simpler than the FLL circuit and it uses a phase-sensitive detector (PSD) which can simply be a multiplier. The input and output signals, after multiplication, are filtered by a narrowband, low-pass loop filter, amplified, and applied to the VCO. It is shown in Appendix I that for small phase differences $(\theta_i - \theta_o)$, i.e. near phase lock, the circuit operation is essentially linear and the output from the loop filter is $v_o \simeq A\dot{\theta}_i(t)$.

Initially, if the input and output signal frequencies and phases are different, the 'error' signal moves the VCO frequency to produce a frequency lock with $\omega_i = \omega_o$ and then moves phase θ_o as close to θ_i as possible to achieve a phase lock. However, for proper operation $(\theta_i - \theta_o)$ must always be finite, but can be made very small if the loop gain is large.

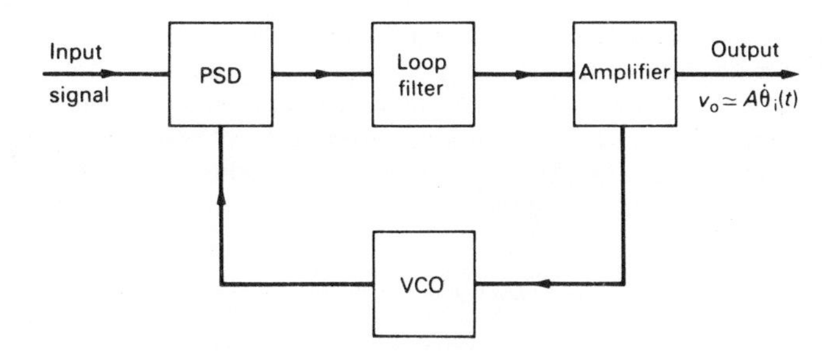

Fig. 6.12

Hence, if the input signal is an FM signal and the loop bandwidth is sufficient to include the baseband frequencies, the control voltage v_c corresponds to the audio modulation and the circuit functions as an FM demodulator. Here, again, as the loop filter bandwidth is much smaller than that of the input FM signal, there is an (S/N) improvement at the output and the FM threshold can be reduced by about 6 dB in practice.

Another application of the PLL is to provide a stable frequency source by cleaning-up phase noise in an oscillator. To achieve this, the VCO is usually the frequency source to be stabilised and the input signal is provided by a very stable crystal oscillator. By means of the loop, the VCO is locked to the frequency of the crystal and any phase noise is considerably reduced by the error signal.

Usually, the frequencies of the VCO and the stable oscillator are not the same but, by using various stages of frequency multiplication and division, the necessary frequency transformation is achieved. Furthermore, a more versatile circuit using this technique is capable of providing a number of different stabilised frequencies and is called a *frequency synthesiser*.

6.6 Tracking loops[39,40]

In communication systems which transmit no reference carrier, e.g. DSBSC, there is a need to recover the carrier in some way in order to re-insert it at the receiver prior to coherent demodulation. This can be achieved using either a Costas loop, a squaring loop, or a remodulation loop.

Costas loop

Here, the phase of the incoming carrier, whose frequency is known beforehand, is extracted from the suppressed carrier signal $s(t)$ plus noise $n(t)$ by multiplying the input voltages in two phase detectors (multipliers) with the output from the VCO and a 90° phase-shifted version of that voltage respectively. The output products are then filtered by low-pass filters and the resultant signals, which are known as the in-phase (I) channel and the quadrature phase (Q) channel, are fed to a phase discriminator, such as a multiplier. The output from the phase discriminator is filtered and the d.c. component remaining is used to shift the local VCO phase by the right amount. Hence, a coherent carrier signal is obtained from the VCO which can be used to demodulate the DSBSC signal. The loop is illustrated in Fig. 6.13.

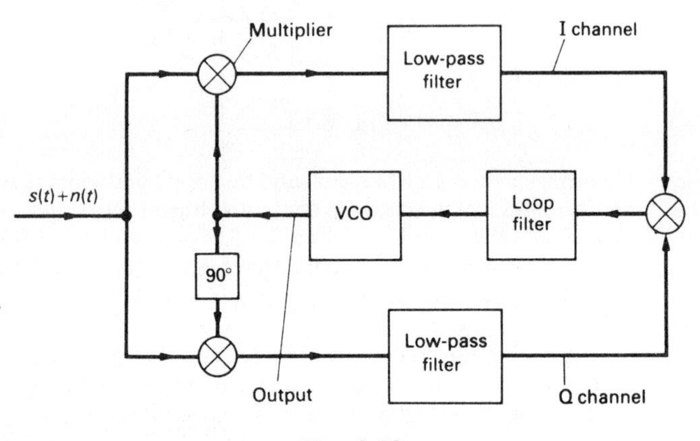

Fig. 6.13

Squaring loop

The incoming signal at frequency ω_c is squared and therefore doubled to $2\omega_c$. With other components, it is filtered by a narrow band-pass filter centred at $2\omega_c$. The output of the filter is subsequently fed into a conventional phase-locked loop which moves the VCO frequency and phase to lock with the input signal. The output of the VCO is the required coherent signal but at double the input frequency. It is therefore divided by a factor of two in a frequency divider which then gives a coherent signal output at frequency ω_c for demodulation purposes. The loop is illustrated in Fig. 6.14.

Remodulation loop

The circuit shown in Fig. 6.15 makes an estimate $\hat{m}(t)$ of the incoming signal $s(t)$ with modulation $m(t)$ using a basic phase-locked loop in which the VCO phase is initially incoherent. After extraction of the modulation estimate $\hat{m}(t)$, it is used to remodulate a suitably delayed version of the incoming signal $s(t)$. This causes the original 180° phase shifts in the incoming signal, e.g. biphase PSK, to be shifted to 360° and so produces an unmodulated carrier signal. This output signal is then used to correct the VCO phase and phase-lock it to the required carrier phase.

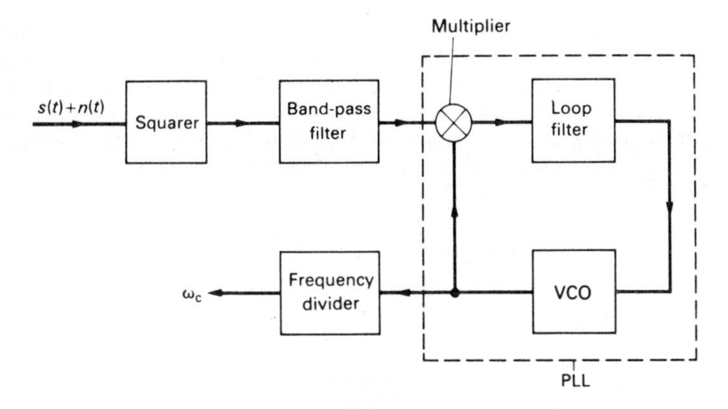

Fig. 6.14

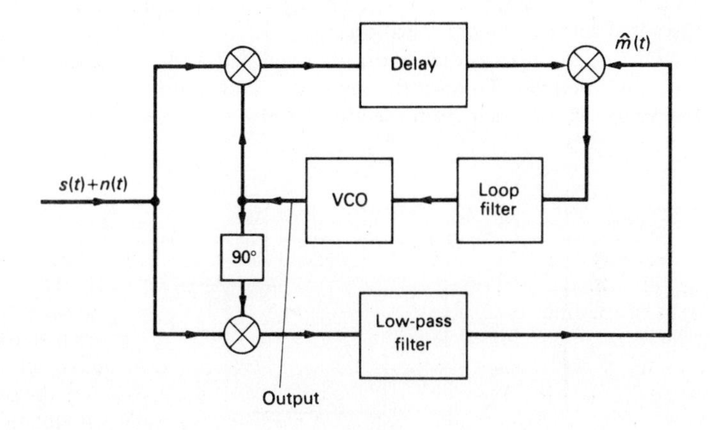

Fig. 6.15

6.7 Pulse demodulation

In the case of pulse signals, the modulated pulse train on reception has to be demodulated in a manner which will reproduce the original modulation. An earlier study of the spectrum of PAM revealed that the modulation could be recovered by passing the modulated pulse train through an appropriate low-pass filter with a cut-off frequency W, where W is the highest modulating frequency component.

In a PAM demodulator, the low-pass filter may be preceded by a sample and hold circuit to increase the output signal. The circuit charges up a capacitor through a low impedance switch *during* a sample pulse. It is then momentarily switched across a high-impedance output for the interpulse interval T_s. The capacitor voltage therefore changes at each sample pulse but remains fairly constant in between pulses. This produces a 'staircase' waveform which is then smoothed by a low-pass filter. The arrangement is shown in Fig. 6.16.

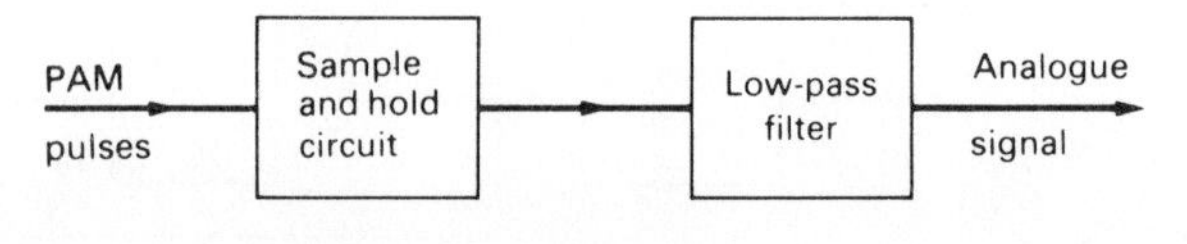

Fig. 6.16

The demodulation of PDM or PPM can also be achieved by using a low-pass filter, if the modulation is small compared to the interpulse period to avoid distortion. However, demodulation can be improved on a pulse-to-pulse basis by using some form of synchronisation.

In a PDM demodulator, each pulse is integrated and then sampled by a sample and hold circuit. To reset the integrator and sampling circuit to zero, synchronisation is obtained from the PDM pulse train and this is shown in Fig. 6.17.

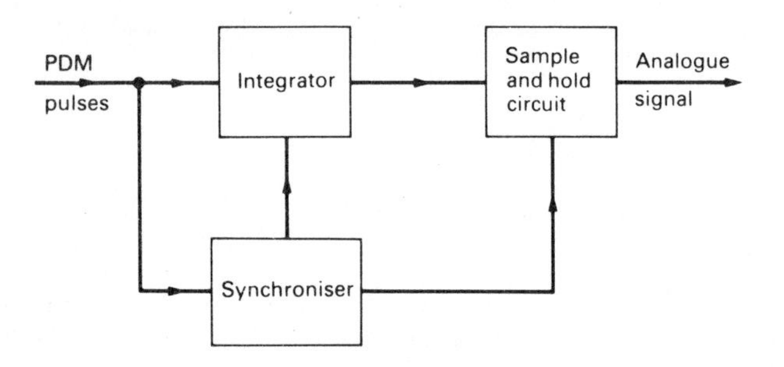

Fig. 6.17

For a PPM demodulator, the PPM pulses can be converted to PDM pulses and then demodulated as described above. The conversion to PDM is achieved by using a *slicer*, to reduce noise effects, and a flip-flop which can be synchronised by a reference pulse from the PPM pulse train as shown in Fig. 6.18. A slicer is a type of limiter which operates at two threshold levels of the input pulse and allows only that part of the pulse between those levels to pass. It produces an output pulse of constant amplitude with less noise on it.

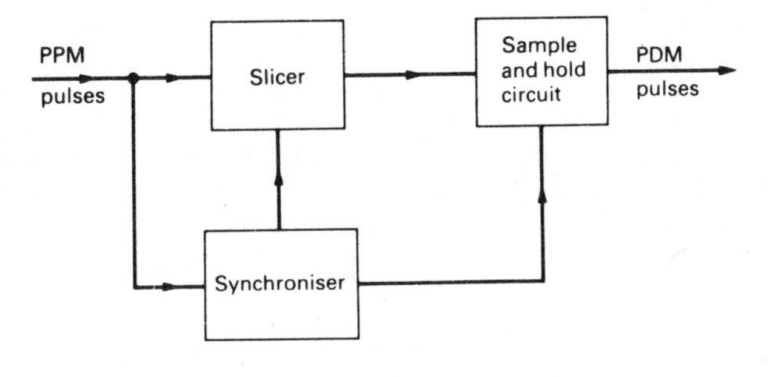

Fig. 6.18

For PCM signals, the coded pulse groups have to be decoded and expanded first to yield the quantised samples. This is achieved by a decoder and expander and the PAM samples are then passed through a low-pass filter to yield the required analogue signal. The technique is illustrated in Fig. 6.19.

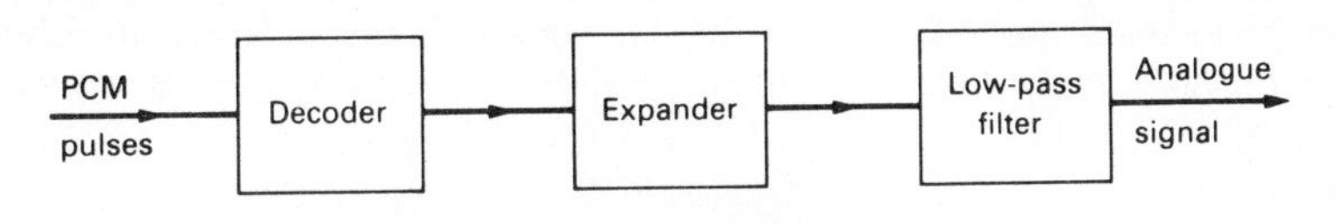

Fig. 6.19

Example 6.4

Explain how band-limiting in a digital communication system may lead to *intersymbol interference*. Assuming that the high frequencies are cut off by a simple *RC* network, define the *crosstalk ratio* and derive it in terms of the network parameters and parameters of the pulse train. If sixteen 2 kHz channels are sampled at the minimum satisfactory rate and multiplexed, calculate the crosstalk ratio between channels for $RC = 8 \times 10^{-7}$ second. (C.E.I)

Solution

In a digital communication system, rectangular or square pulses may be used to represent the symbols 0 or 1 and the sharp pulse edges require a very large bandwidth. However, bandwidth is usually restricted in communication systems to avoid interference between radio stations or because of coaxial cable limitations.

Hence, the pulses are usually band-limited in transmission by suitable design or filter characteristics. This band-limiting broadens each pulse and, at the receiver, it causes some overlap between adjacent pulses as shown in Fig. 6.20. The overlap between adjacent pulses is known as intersymbol interference or crosstalk.

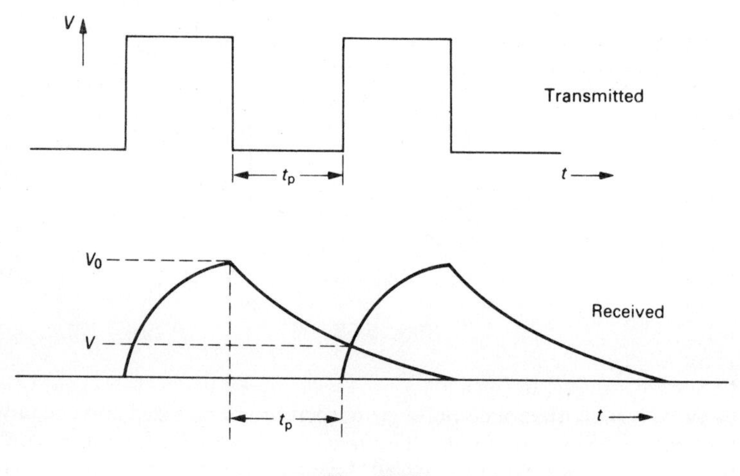

Fig. 6.20

If the high frequencies are cut off by a simple RC network, for a peak pulse voltage V_0 and an interpulse period t_p, the overlap will occur at a voltage V. Hence, we have

$$V = V_0 e^{-t_p/RC}$$

or

$$\frac{V_0}{V} = e^{t_p/RC}$$

The crosstalk ratio is defined in dB as $8{\cdot}686 \ln (V_0/V)$ and so we obtain

$$\text{crosstalk ratio} = \frac{8{\cdot}686 t_p}{RC} \text{ dB}$$

Problem
Minimum sampling frequency per channel = 2 × 2 kHz = 4 kHz
Sampling frequency for 16 channels = 16 × 4 kHz = 64 kHz

For simplicity, assume square pulses and a duty ratio of 1/2. Hence, we obtain

$$2t_p = 1/(64 \times 10^3)$$

or

$$t_p = \frac{1}{(128 \times 10^3)} \simeq 8 \times 10^{-6} \text{ s}$$

For $RC = 8 \times 10^{-7}$ s and $t_p = 8 \times 10^{-6}$ s, we obtain

$$\text{crosstalk ratio} = \frac{8{\cdot}686 \times 8 \times 10^{-6}}{8 \times 10^{-7}} \simeq 87 \text{ dB}$$

Comment
A typical value for this ratio is 60 dB and so the sampling frequency per channel could be increased to 5 kHz in practice. This would yield a crosstalk ratio of about 68 dB.

6.8 Digital demodulation[3, 41]

The various types of digital modulation used in practice are frequency-shift-keying (FSK), phase-shift-keying (PSK), differential phase-shift-keying (DPSK), and quadrature phase-shift-keying (QPSK), as described earlier in Section 5.11.

To demodulate the digital data at the receiver, either coherent or non-coherent techniques may be employed and will usually involve some form of matched filtering or correlation detection, together with a threshold decision after each bit interval T. The threshold level is optimally set at zero volts for signal outputs which may be either positive or negative.

Coherent demodulation

An FSK demodulator using matched filter detection employs two matched filters, each tuned to frequencies f_1 and f_2 corresponding to either a mark or space respectively. The outputs from the filters are subtracted and a threshold decision is made at a time T after each bit. A positive output signifies a mark and a negative output signifies a space as illustrated in Fig. 6.21(a).

An FSK demodulator employing correlation detection uses coherent signal frequencies f_1 and f_2 at the receiver together with a multiplier and integrator.

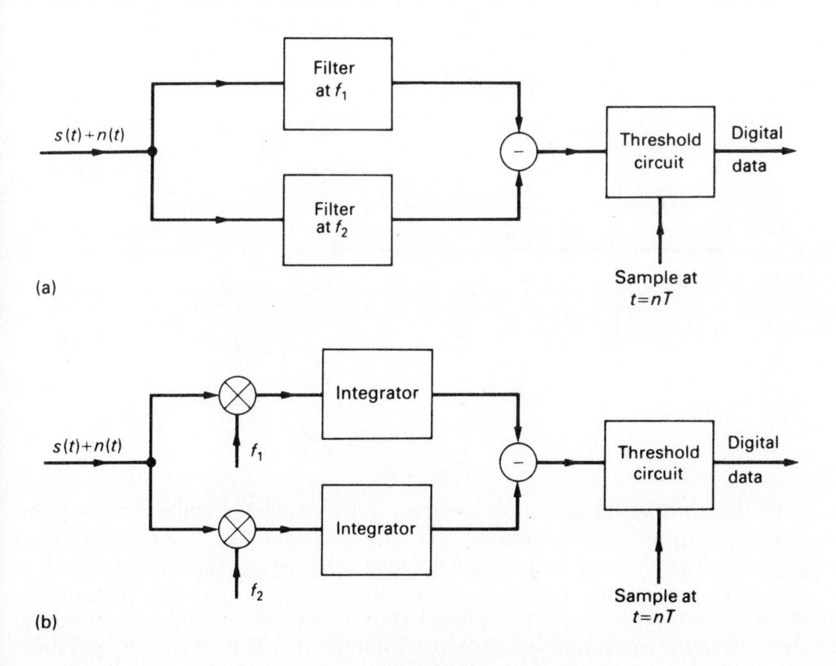

Fig. 6.21

The outputs from the integrators are subtracted and a threshold decision at time T after each bit determines whether a mark or space has been transmitted as illustrated in Fig. 6.21(b). In both these methods it is necessary to extract bit timing at the receiver, from the digital stream, to operate the decision circuit at the optimum time T after each bit interval. Furthermore, the integrators must be reset to zero after each bit of information.

PSK demodulators use matched filter or correlation techniques similar to those employed for FSK. However, the matched filter may be difficult to implement, while precise sampling is difficult to achieve and the distortion of received pulses, due to channel frequency response, may give rise to intersymbol interference at the filter output. The matched filters may be typically transversal filters or surface acoustic-wave devices.

As correlation detection involve the multiplication of in-phase or out-of-phase signals, only one type of reference signal is required at the receiver, which corresponds to a mark signal when there is no phase reversal of the carrier. As before, the threshold of the decision circuit determines whether a mark or space has been transmitted. The circuits are illustrated in Fig. 6.22.

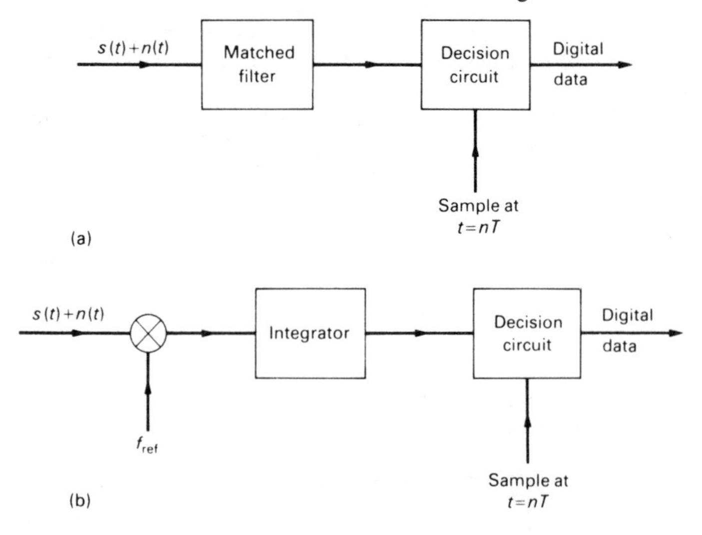

Fig. 6.22

In the correlation detectors used for FSK or PSK demodulation, good synchronisation between received and reference signals is required. A relative phase error of 25° will give an effective signal-to-noise ratio loss of about 1 dB in comparison with the perfectly synchronised system. The position of the decision sampling pulse is less critical than with band-pass matched filtering, since the integrator output ideally has a linear increase in voltage up to this time.

The integrator is then 'dumped' or reset to zero after each decision and so it is often called an 'integrate and dump' filter.

In a DPSK demodulator, each bit is used as a reference bit for the following bit and so it is stored at the receiver by means of a delay circuit equal to the bit time T. This is followed by a correlation detector and a decision circuit as previously described. A typical circuit arrangement is shown in Fig. 6.23.

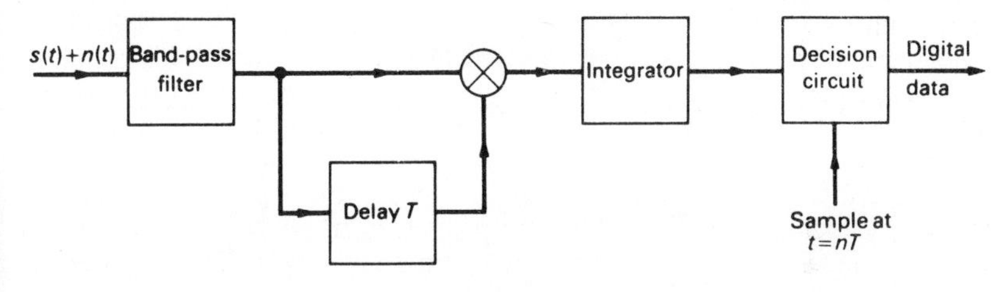

Fig. 6.23

In a QPSK demodulator, a correlation type of receiver is also used. Each of the four transmitted signals may have its own synchronised reference at the receiver, and the receiver is independently correlated with each of these references. The correlator giving the largest output at the end of the integration period defines the received signal.

An alternative detector configuration is shown in Fig. 6.24. Each received signal pulse is resolved into two quadrature components and the polarities of these components define the received data bits. The signal output from each branch is reduced by $1/\sqrt{2}$ compared to the output of the biphase PSK correlation receiver described earlier. Thus, for a given bit rate, an increase of 3 dB in signal-to-noise ratio is required to maintain the same bit error rate as for the biphase system. The signal outputs are shown in Table 6.1.

Non-coherent demodulation

Most practical circuits use non-coherent FSK demodulation as shown in Fig. 6.25. Two band-pass filters are used whose bandwidths are approximately equal to the reciprocal of the bit rate. The filter outputs are envelope-detected and the envelope amplitudes are compared after the reception of each signal pulse.

The optimum non-coherent receiver uses matched filters followed by envelope detectors. This detector is then matched to the signal envelopes and not to the signals themselves. The phase of the signal carrier is of no importance in defining the envelope and hence no phase information is used in non-coherent detection. The penalty for using conventional band-pass filters as opposed to matched filters is about 1–2 dB in signal-to-noise ratio, but the instrumentation is simplified.

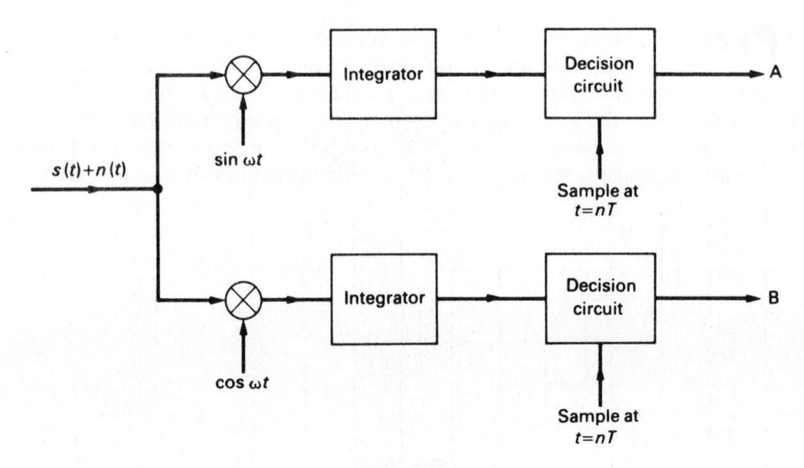

Fig. 6.24

Table 6.1

Input pulse	*Output A*	*Output B*
$\sin(\omega t + 45°)$	1	1
$\sin(\omega t + 135°)$	0	1
$\sin(\omega t + 225°)$	0	0
$\sin(\omega t + 315°)$	1	0

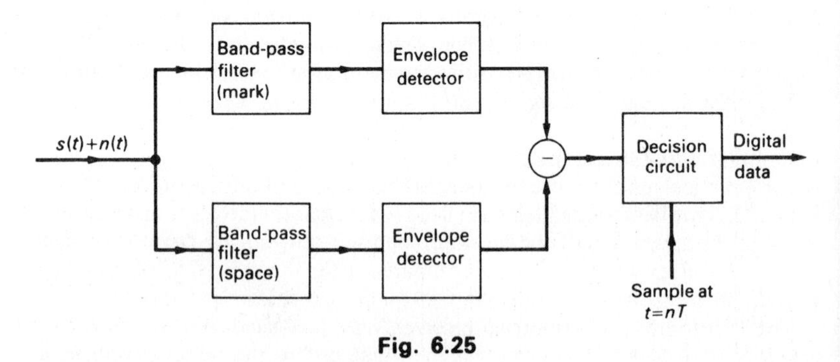

Fig. 6.25

Comments

1. Coherent DPSK and non-coherent FSK are popular data transmission systems. DPSK is instrumented with greater difficulty and is generally preferred when bandwidth is to be conserved.
2. Details of the bit error rate (BER) are given elsewhere.[3]

Problems

1 State the merits of single-sideband transmission and show, with the aid of diagrams, how it is applied to multichannel line telephony. Explain why several modulator stages are generally used.

In a certain radio transmitter the depth of modulation is 100 %. The carrier component level is reduced by 20 dB after the modulator. The reduced carrier and one of the sidebands are subsequently connected to a power amplifier. If the maximum power input to this amplifier is limited to 2 W, determine the level of the original carrier. (U.L.)

2 The power output of an AM anode-modulated transmitter is 1 kW. The efficiency of the final stage is 70 %, the modulation depth is 0·5, and the efficiency of the modulator is 55 %. Calculate

(a) the modulation power required at the anode of the final stage,

(b) the anode dissipation of the modulator stage.

3 Describe, using sketches when necessary, how the amplitude and frequency of a modulating signal are conveyed by

(a) amplitude modulation,

(b) frequency modulation.

Discuss briefly the advantages and disadvantages of FM compared with AM in a VHF communication system. The RF bandwidth of an FM transmitter is 80 kHz when a 6 kHz modulating signal is applied to the input. What bandwidth is required if the modulating signal level is reduced by 6 dB? (C.G.L.I)

4 Compare and contrast a narrowband frequency-modulated waveform with an amplitude-modulated waveform having the same carrier power.

The modulation in an FM transmitter is achieved by variation of the tuning capacitance of an oscillator operating at a mean frequency of 3 MHz. The coil used in the parallel tuned circuit has an inductance of 10 μH. If the modulated waveform is frequency-multiplied to give an output of 60 MHz with a maximum frequency deviation of 200 kHz, determine the change in value of capacitance to be produced by the modulating signal. (U.L.)

5 State the meaning of the following terms when applied to a frequency-modulated transmission

(a) modulation index,

(b) frequency deviation,

(c) deviation ratio.

When the modulation index of a certain FM transmitter is 6, and the bandwidth in use is 140 kHz, what is its frequency deviation?

Explain the effect of altering the deviation in an FM system on the output from a receiver due to an interfering signal whose frequency is close to that of the carrier of the wanted signal. (C.G.L.I.)

6 With the aid of diagrams, describe in detail the principle of the varactor diode. Give an approximate expression for the relationship between capacitance and applied voltage in this diode. State the order of capacitance variation commonly obtainable. (C.G.L.I.)

7 Show that a frequency-modulation communication system can give an improved output signal-to-noise ratio compared with an amplitude-modulation system. Specify carefully the conditions of comparison and the assumptions made. Sketch the output signal-to-noise ratio as a function of the input signal-to-noise ratio in each case and account for any special features of the curves. (C.E.I.)

8 (a) Explain the difference between frequency modulation and phase modulation.

(b) When a carrier is phase-modulated by a modulating signal of 1 kHz, the phase deviation is 5 radians. If the amplitude of the modulating signal remains constant, but its frequency is changed to 5 kHz, calculate

(i) the phase deviation of the carrier,

(ii) the frequency deviation of the carrier.

(c) The voltage of a frequency-modulated wave is

$$V = 5 \sin (2\pi\, 10^8 t - 20 \cos 4\pi\, 10^3 t) \text{ volts}$$

Explain how the practical bandwidth is determined and estimate its value. (C.G.L.I.)

9 Discuss the relationship between phase modulation and frequency modulation of a carrier wave, and derive expressions for the instantaneous values of phase- and frequency-modulated signals, where the modulating signal has the form $V_m \cos \omega_m t$. A signal $v(t) = 0{\cdot}1 \cos (2\pi\, 10t)$ is used to frequency-modulate a 1 MHz carrier. The peak frequency shift caused by the 0·1 V modulating voltage is 100 Hz. What is the required receiver bandwidth?

Justify any assumptions you make and explain why the same method of calculation would be invalid for a modulating frequency of 500 Hz. (U.L.)

10 Give a block diagram of a system for the generation of sampled data from an analogue baseband signal. If the highest baseband frequency is 15 kHz, suggest with reasons a suitable duration for each sample. Prove that the sample rate must be at least 30 kHz and explain the form of distortion known as *aliasing*.

Give the principles of two interpolation techniques for recovery of the analogue signal from the sampled information and sketch circuits to implement each method. (C.E.I.)

11 Describe briefly the basic principles of the time division multiplex system of transmission.

State the effect on interchannel crosstalk in a pulse system of

(a) attenuation distortion,

(b) phase distortion.

Several speech channels are to be transmitted over a radio link by time division pulse position modulation. Using the information given below, estimate the maximum number of channels which may be transmitted.

Channel sampling frequency	8 kHz
Pulse width	2 μs
Pulse shift	±3 μs
Minimum adjacent channel pulse separation	2 μs

(U.L.)

12 Explain in detail the principles of pulse code modulation. Describe a method whereby pulse code signals may be generated from analogue signals for transmission over a communication network and also translated back to their original analogue form at the distant end.

Discuss whether any advantage would be gained from the use of a slicer or a peak limiter in a PCM receiver. (C.G.L.I.)

13 Twelve 4 kHz channels are to be multiplexed and transmitted on a PCM link. The rms signal-to-quantising noise level is to be better than 30 dB and linear quantising is in use. Give a simple block diagram of the modulation system, and estimate the minimum bandwidth. Why is the above quantising method unrealistic when applied to speech? How, in practice, may speech be efficiently quantised? (C.E.I.)

14 Discuss the merits of delta modulation, explaining carefully how it differs from pulse modulation.

Draw a block diagram of a delta-modulation system and describe the function of each component part. (C.G.L.I.)

15 Derive an expression for the signal-to-quantisation noise of a delta-modulation system in terms of the input signal $f(t)$, the highest modulating frequency f_m, and the sampling frequency f_s. State any assumptions made. Hence, evaluate the ratio for the case of a speech signal with $f_m = 3{\cdot}4$ kHz and $f_s = 32$ kHz.

16 Draw a block diagram for a superheterodyne radio receiver. Discuss briefly the factors that influence the choice of the intermediate frequency.

Design a diode envelope detector for distortionless operation with a 500 kHz carrier modulated to a maximum depth of 80% by frequencies in the range of 50 Hz to 10 kHz.

Draw a circuit for the detector showing how it is coupled to the first AF amplifier stage. Allocate values to all significant components and check the limiting condition for peak clipping. (C.E.I.)

17 Envelope modulation is preferred to single-sideband transmission for national broadcasting. Give the arguments for and against this, using elementary analysis where necessary. In the light of the rapid development of integrated circuits, are there sound technical reasons for expecting a change in the situation? (C.E.I.)

18 An amplitude-modulated signal is transmitted with a carrier power of 100 kW and 18 kW sideband power. The audio bandwidth of an AM communication receiver is 4 kHz and the mean noise power received per unit bandwidth is 10^{-3} W/Hz. What is the signal-to-noise ratio at the output of an envelope detector? Assume a sinusoidal modulating signal.
How would the ratio change if a DSBSC transmission were used? Assume equal transmitter power for both systems.

19 A signal of the form $f(t) = 10 \sin 2000\, t$ is transmitted by means of double-sideband suppressed carrier modulation (DSBSC). During transmission, Gaussian white noise power with a spectral density of 10^{-3} W/Hz is added to the signal. If synchronous demodulation is used at the receiver, determine the signal-to-noise ratio at the output of the demodulator. Assume 100% modulation at the transmitter.

20 Sketch the relationships between output and input signal-to-noise ratios for *coherent* and *linear* detectors, explaining the difference between them. Detail the other main differences between the circuits, and show how these affect the different fields of application. (C.E.I.)

21 For a certain error probability, the decibel values of the energy per bit to noise power spectral density are 9·5 and 12·5 for coherent PSK and coherent FSK respectively. Hence, using the approximation

$$\operatorname{erfc}(x) \simeq \frac{1}{\sqrt{\pi}\, x} e^{-x^2}$$

determine the corresponding values for
(a) DPSK,
(b) non-coherent FSK.

Answers

1 7·7 W
2 159 W, 130 W
3 46 kHz
4 3·75 pF
5 60 kHz
The interfering signal output is reduced if the deviation is increased.
6 $C \propto 1/\sqrt{V}$
7 5 pF to 15 pF per volt typically
8 1 radian, 5 kHz, 84 kHz
9 $B = 2(\Delta f + f_m) = 220\,\text{Hz}$ (wideband FM)
$B = 2f_m = 1\,\text{kHz}$ (narrowband FM)
10 $2\,\mu\text{s}$ for 8-bit PCM code and $(S/N_q) > 30\,\text{dB}$
In a sampled-data signal, sideband components near the sampling frequency 'fold-over' and overlap the baseband frequencies if the sampling frequency is too low. These *aliasing* components cause distortion.
Two interpolation techniques used are direct low-pass filtering or a sample and hold filter circuit.
11 12 channels
12 The use of a slicer or peak limiter will improve the output (S/N) ratio and reduce digital errors.
13 336 kHz for 672 kbit/s
Speech is made up of low amplitudes for most of the time. A more realistic method is to use non-linear quantisation.
15 $\dfrac{S}{N_q} = \dfrac{3f_s}{\sigma^2 f_m} \times \overline{f^2(t)}$ where σ is the step size.

$\dfrac{S}{N_q} = \dfrac{3}{4\pi^2}(f_s/f_m)^3 \simeq 18\,\text{dB}$ for a sinusoidal input.

18 36 dB, 8 dB improvement
19 25 dB
21 10·2 dB, 13·4 dB

References

1 KENNEDY, G. *Electronic Communication Systems*, McGraw-Hill (1977).
2 PEEBLES, P. Z. *Communication System Principles*, Chapter 7. Addison-Wesley (1976).
3 CONNOR, F. R. *Noise*, Chapter 6. Edward Arnold (1982).
4 DIXON, R. C. (ed) *Spread Spectrum Techniques*. IEEE Press (1976).
5 GOLOMB, S. W. *Digital Communications With Space Applications*. Prentice-Hall (1964).
6 CLARKE, K. K. and HESS, D. T. *Communication Circuits, Analysis and Design*. Addison-Wesley (1971).
7 PAPPENFUS, E. W. *et al. Single Sideband Principles and Circuits*. McGraw-Hill (1964).
8 *Proceedings Institute of Radio Engineers*, **44**, 1661, 1956. (Single Sideband Issue.)
9 TURNER, L. W. *Electronic Engineer's Reference Book*. Newes-Butterworth (1976).
10 BRAY, W. J. and MORRIS, D. W. Single-Sideband Multichannel Operation of Short-Wave Point-to-Point Radio Links. *Post Office Electrical Engineers Journal*, **45**, 97, October 1952.
11 FINK, D. G. *Electronics Engineers' Handbook*. McGraw-Hill (1975).
12 CARSON, J. R. Notes on the Theory of Modulation. *Proceedings Institute of Electrical and Electronic Engineers*, **51**, 893, 1963.
13 TAUB, H. and SCHILLING, D. L. *Principles of Communications*. McGraw-Hill (1971).
14 COOK, A. B. and LIFF, A. A. *Frequency Modulation Receivers*. Prentice-Hall (1968).
15 PANTER, P. F. *Modulation, Noise and Spectral Analysis*. McGraw-Hill (1965).
16 MANDL, M. *Fundamentals of Electronics*, Prentice-Hall (1973).
17 ARMSTRONG, H. A Method of Reducing Disturbances in Radio Signalling by a System of Frequency Modulation. *Proceedings Institute of Radio Engineers*, **24**, 689, May 1936.
18 CONNOR, F. R. *Signals*. Edward Arnold (1982).
19 FITCH, E. The Spectrum of Modulated Pulses. *Journal Institute of Electrical Engineers*, **94**, Part 3A, 556, 1947.
20 OLIVER, B. M. *et al.* The Philosophy of PCM. *Proceedings Institute of Radio Engineers*, **36**, 1324, November 1948.
21 CATTERMOLE, K. W. *Principles of Pulse Code Modulation*. Iliffe (1969).
22 HAYKIN, S. *Communication Systems*, Chapter 6. John Wiley (1978).
23 SCHOUTEN, J. F. *et al.* Delta Modulation. *Philips Technical Review*, **13**, 237, March 1952.
24 ABATE, J. E. Linear and Adaptive Delta Modulation. *Proceedings Institute of Electrical and Electronic Engineers*, **55**, 298–308, March 1967.
25 INOSI, H. and YASUDA, Y. A Unity Bit Coding Method by Negative Feedback. *Proceedings Institute of Electrical and Electronic Engineers*, 1524, November 1963.
26 DEVEREUX, V. G. Application of PCM to Broadcast Quality Video Signals. *The Radio and Electronics Engineer*, **44**, 373–81, July 1974.
27 JAYANT, N. S. Digital Coding of Speech Waveforms, PCM; DPCM and DM

Quantizers. *Proceedings Institute of Electrical and Electronic Engineers*, **62**, 611–32, May 1974.
28 CONNOR, F. R. *Noise*, Chapter 3. Edward Arnold (1982).
29 DIXON, R. C. *Spread Spectrum Systems*. John Wiley (1976).
30 BAIER, P. W. and PANDIT, M. Spread Spectrum Communication Systems. *Advances in Electronics and Electron Physics*, **53**, 209–67, 1980.
31 HAYKIN, S. *Communication Systems*, Chapter 5. John Wiley (1978).
32 PEEBLES, P. Z. *Communication System Principles*, Chapter 5. Addison-Wesley (1976).
33 FOSTER, D. E. and SEELEY, S. W. Automatic Tuning, Simplified Circuits and Design Practice. *Proceedings Institute of Radio Engineers*, **25**, 289, 1937.
34 SEELEY, S. W. The Ratio Detector. *RCA Review*, **8**, 201, 1947.
35 WOBSCALL, D. *Circuit Design for Electronic Instrumentation*. McGraw-Hill (1979).
36 KLAPPER, J. and FRANKLE, J. T. *Phase-Locked and Frequency-Feedback Systems*. Academic Press (1972).
37 RICE, S. O. Noise in FM Receivers. *Proceedings of the Symposium on Time Series Analysis*. Chapter 25, 395–424. John Wiley (1963).
38 GARDNER, F. M. *Phase Lock Techniques*. John Wiley (1979).
39 LINSEY, W. C. and SIMON, M. K. *Telecommunication Systems Engineering*. Prentice-Hall (1973).
40 DIDDAY, R. L. and LINSEY, W. C. Subcarrier Tracking Methods. *Institute of Electrical and Electronic Engineers Transactions*, **Com-16**, 541–50, August 1968.
41 PEEBLES, P. Z. *Communication System Principles*, Chapter 8. Addison-Wesley (1976).
42 SCHWARTZ, M. *et al. Communication Systems and Techniques*. McGraw-Hill (1966).
43 GOLOMB, S. W. *et al. Shift Register Sequences*. Holden-Day (1967).

Appendices

Appendix A: Hilbert transform[42]

If $s(t)$ is any real modulating signal, let the *Hilbert transform* of $s(t)$ be denoted by $\hat{s}(t)$. It is given by the expression

$$\hat{s}(t) = \frac{1}{\pi} \int_{-\infty}^{+\infty} \frac{s(\tau)}{(t-\tau)} \, d\tau$$

where τ is any arbitrary delay variable and $\hat{s}(t)$ is obtained physically by shifting all the frequency components of $s(t)$ through 90°, using a wide-band phase-shift network known as a *Hilbert transformer*.

The expression for $\hat{s}(t)$ also represents the time convolution of the function $1/\pi t$ with $s(t)$ and may be written as

$$\hat{s}(t) = \frac{1}{\pi t} * s(t)$$

where the asterisk sign $*$ denotes convolution. Hence, if $F(\omega)$ and $\hat{F}(\omega)$ are the Fourier transforms of $s(t)$ and $\hat{s}(t)$ respectively, convolution in the time domain is equivalent to multiplication in the frequency domain and so we obtain

$$\hat{F}(\omega) = -\mathrm{j}\,\mathrm{sgn}(\omega)\, F(\omega)$$

where $-\mathrm{j}\,\mathrm{sgn}(\omega)$ is the Fourier transform of $1/\pi t$ and $\mathrm{sgn}(\omega)$ is the *signum* function defined by

$$\begin{aligned} f(\omega) &= \mathrm{sgn}(\omega) = 1 \qquad && \omega > 0 \\ f(\omega) &= \mathrm{sgn}(\omega) = 0 && \omega = 0 \\ f(\omega) &= \mathrm{sgn}(\omega) = -1 && \omega < 0 \end{aligned}$$

as illustrated in Fig. A.1.

Analytic signal

The Hilbert transform can be used to represent an *analytic* signal $S(t)$ defined by

$$S(t) = s(t) + \mathrm{j}\,\hat{s}(t)$$

or

$$S(t) = |S(t)|\,\mathrm{e}^{\mathrm{j}\phi(t)}$$

where $|S(t)| = \sqrt{s^2(t) + \hat{s}^2(t)}$ and $\phi(t) = \tan^{-1}[\hat{s}(t)/s(t)]$

with $s(t) = \mathrm{Re}[S(t)]$ and $\hat{s}(t) = \mathrm{Im}[S(t)]$

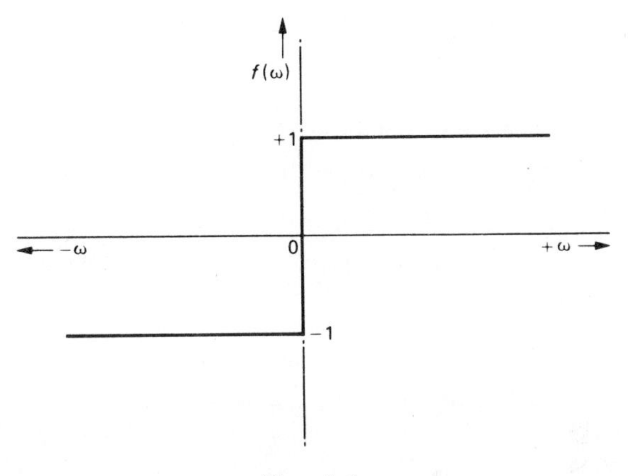

Fig. A.1

The analytic signal is a complex signal representation, with its real and imaginary parts forming a Hilbert transform pair. It is useful in the study of modulation theory, e.g. the analysis of an SSBSC signal.

SSBSC signal

If $s(t)$ is a modulating signal and $\sin \omega_c t$ is any carrier, the phase-shift method of generating an SSBSC signal yields an output $v_o(t)$ given by

$$v_o(t) = s(t) \sin \omega_c t + \hat{s}(t) \cos \omega_c t$$

where $\hat{s}(t) = \mathrm{j}\, s(t)$ and is due to the 90° phase-shift network. Hence, if $s(t) = \sin \omega_m t$ is a single-tone modulating signal, $\hat{s}(t) = \cos \omega_m t$ and we obtain

$$v_o(t) = \sin \omega_c t \sin \omega_m t + \cos \omega_c t \cos \omega_m t$$

or

$$v_o(t) = \cos (\omega_c - \omega_m)t$$

which is an SSBSC signal (lower sideband). Hence, an SSBSC signal can be easily represented in terms of the analytic signal $S(t)$ by the expressions

$$v_o(t) = \mathrm{Re}[S(t) \mathrm{e}^{\mathrm{j}\omega_c t}]$$

or

$$v_o(t) = \mathrm{Re}[|S(t)| \mathrm{e}^{\mathrm{j}\{\omega_c t \pm \phi(t)\}}]$$

assuming the carrier and modulating signals are *cosine* functions.

Note

This representation involves the use of both amplitude and phase information, i.e. $|S(t)|$ determines the form of the signal envelope and $\phi(t)$ involves the phase of the signal. Hence the need for accurate phase and frequency information at the receiver if the demodulation is to be correct. Furthermore, the expression for $v_o(t)$ takes on the form of a 'rotating phasor'.

Appendix B: Frequency modulation

An FM carrier wave can be represented by

$$v_c = V_c \sin[\omega_c t - m_f \cos \omega_m t]$$

or

$$v_c = V_c[\sin \omega_c t \cos(m_f \cos \omega_m t) - \cos \omega_c t \sin(m_f \cos \omega_m t)]$$

Furthermore, it can be shown that

$$\cos(m_f \cos \omega_m t) = J_0(m_f) - 2J_2(m_f)\cos 2\omega_m t + 2J_4(m_f)\cos 4\omega_m t - \ldots$$
$$\sin(m_f \cos \omega_m t) = 2J_1(m_f)\cos \omega_m t - 2J_3(m_f)\cos 3\omega_m t + \ldots$$

Hence

$$\begin{aligned} v_c &= V_c[\sin \omega_c t\{J_0(m_f) - 2J_2(m_f)\cos 2\omega_m t + 2J_4(m_f)\cos 4\omega_m t + \ldots\} \\ &\quad - \cos \omega_c t\{2J_1(m_f)\cos \omega_m t - 2J_3(m_f)\cos 3\omega_m t + \ldots\}] \\ &= V_c[J_0(m_f)\sin \omega_c t - 2J_1(m_f)\cos \omega_c t \cos \omega_m t - 2J_2(m_f)\sin \omega_c t \cos 2\omega_m t \\ &\quad + 2J_3(m_f)\cos \omega_c t \cos 3\omega_m t + 2J_4(m_f)\sin \omega_c t \cos 4\omega_m t + \ldots] \end{aligned}$$

Since $\quad 2\cos A \cos B = \cos(A+B) + \cos(A-B)$

and $\quad 2\sin A \cos B = \sin(A+B) + \sin(A-B)$

we then obtain

$$\begin{aligned} v_c = V_c[J_0(m_f)\sin \omega_c t &- J_1(m_f)\{\cos(\omega_c + \omega_m)t + \cos(\omega_c - \omega_m)t\} \\ &- J_2(m_f)\{\sin(\omega_c + 2\omega_m)t + \sin(\omega_c - 2\omega_m)t\} \\ &+ J_3(m_f)\{\cos(\omega_c + 3\omega_m)t + \cos(\omega_c - 3\omega_m)t\} \\ &+ \ldots] \end{aligned}$$

for a modulating signal $v_m = V_m \sin \omega_m t$.

Comments

1. It will be observed that the pairs of sidebands are alternately in quadrature or in phase with the carrier.
2. For a modulating signal $v_m = V_m \cos \omega_m t$, it can be shown that the FM carrier wave is given by

$$\begin{aligned} v_c = V_c[J_0(m_f)\sin \omega_c t &+ J_1(m_f)\{\sin(\omega_c + \omega_m)t - \sin(\omega_c - \omega_m)t\} \\ &+ J_2(m_f)\{\sin(\omega_c + 2\omega_m)t + \sin(\omega_c - 2\omega_m)t\} + \ldots] \end{aligned}$$

Typical plots of the Bessel functions $J_0(m_f)$, $J_1(m_f)$, $J_2(m_f)$, etc. for various values of m_f are shown in Fig. A.2 and values of some useful Bessel functions are given in Table A.1.

Table A.1

m	$J_0(m)$	$J_1(m)$	$J_2(m)$	$J_3(m)$	$J_4(m)$	$J_5(m)$	$J_6(m)$	$j_7(m)$	$J_8(m)$	$J_9(m)$	$J_{10}(m)$
0	1·000	—	—	—	—	—	—	—	—	—	—
0·2	0·990	0·099	0·005	—	—	—	—	—	—	—	—
0·4	0·960	0·196	0·019	0·001	—	—	—	—	—	—	—
0·6	0·912	0·286	0·043	0·004	—	—	—	—	—	—	—
0·8	0·846	0·368	0·075	0·010	0·001	—	—	—	—	—	—
1·0	0·765	0·440	0·114	0·019	0·002	—	—	—	—	—	—
2·0	0·223	0·576	0·352	0·128	0·034	0·007	0·001	—	—	—	—
3·0	−0·260	0·339	0·486	0·309	0·132	0·043	0·011	0·002	—	—	—
4·0	−0·397	−0·066	0·364	0·430	0·281	0·132	0·049	0·015	0·004	—	—
5·0	−0·177	−0·327	0·046	0·364	0·391	0·261	0·131	0·053	0·018	0·005	0·001
6·0	0·150	−0·276	−0·242	0·114	0·357	0·362	0·245	0·129	0·056	0·021	0·006
7·0	0·300	−0·004	−0·301	−0·167	0·157	0·347	0·339	0·233	0·128	0·058	0·023
8·0	0·171	0·234	−0·113	−0·291	−0·105	0·185	0·337	0·320	0·223	0·126	0·060
9·0	−0·090	0·245	0·144	−0·180	−0·265	−0·055	0·204	0·327	0·305	0·214	0·124
10·0	−0·245	0·045	0·254	0·058	−0·219	−0·234	−0·014	0·216	0·317	0·291	0·207

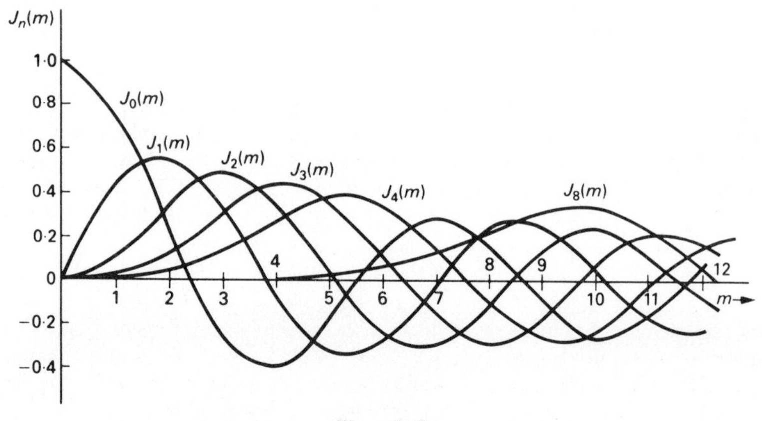

Fig. A.2

Appendix C: Pre-emphasis and de-emphasis

The use of pre-emphasis at the transmitter and de-emphasis at the receiver reduces the noise received for both AM and FM systems, but there is a greater reduction for FM. A typical de-emphasis network is shown in Fig. A.3(a) and its transfer function is given by

$$\frac{v_o}{v_i} = \frac{1/j\omega C}{R + 1/j\omega C} = \frac{1}{1 + j\omega RC}$$

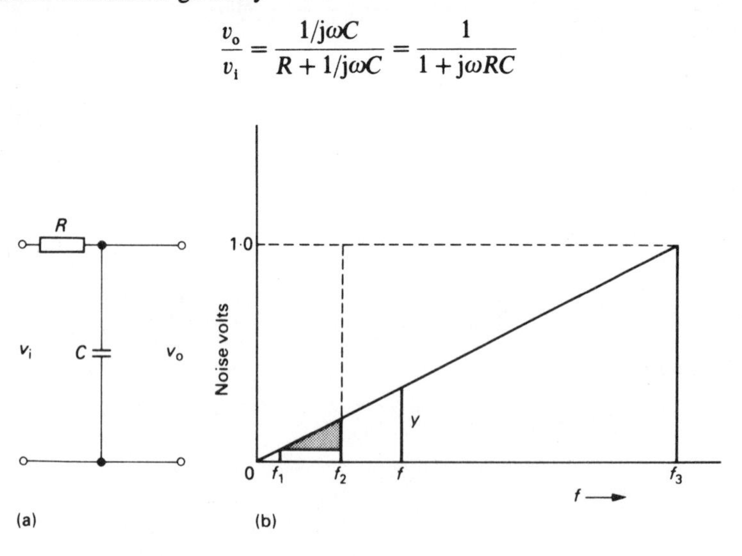

Fig. A.3

or $$|v_o| = \frac{|v_i|}{\sqrt{1+\omega^2 R^2 C^2}}$$

A de-emphasis of $-6\,\text{dB/octave}$ is obtained from a frequency $f_1 = \omega_1/2\pi$ when $\omega_1 RC = 1$. Usually, $RC = 50\,\mu\text{s}$ and so $f_1 = 3{\cdot}1\,\text{kHz}$ and all noise from $f_1 = 3{\cdot}1\,\text{kHz}$ to $f_2 = 15\,\text{kHz}$ is de-emphasised. The shaded area shown in Fig. A.3(b) is the de-emphasised noise for FM and a corresponding effect occurs for AM but this is not shown in Fig. A.3(b). Normalising $v_i = 1$ volt then yields

$$|v_o|^2 = \frac{1}{1+\omega^2 R^2 C^2} = \frac{1}{1+\omega^2/\omega_1^2} = \frac{1}{1+(f/f_1)^2}$$

and the effect of de-emphasis is to reduce the output noise power $|v_o|^2$ by the factor $1/[1+(f/f_1)^2]$ for both AM and FM, assuming the load resistor $R_L = 1\,\Omega$. The new ratio of noise powers can be calculated assuming $f_1 = 3{\cdot}1\,\text{kHz}$, $f_2 = 15\,\text{kHz}$, and $f_3 = 75\,\text{kHz}$.

AM system

Since noise power is proportional to the square of the noise voltage, when $R = 1\,\Omega$ we obtain

$$\text{AM noise power} = \int_0^{f_2} y^2\,\mathrm{d}f = \int_0^{f_2} \frac{\mathrm{d}f}{1+(f/f_1)^2} \qquad (y = 1 \text{ volt})$$

To integrate this expression, let $f/f_1 = x$ or $f = xf_1$ and $\mathrm{d}f = f_1\,\mathrm{d}x$. Hence, we obtain

$$\text{AM noise power} = f_1 \int_0^{f_2} \frac{\mathrm{d}x}{x^2+1} = f_1\,[\tan^{-1} x]_0^{f_2} = f_1 \tan^{-1}\frac{f_2}{f_1} = \frac{\pi f_1}{2}$$

since $\tan^{-1} f_2/f_1 \simeq \pi/2$ if $f_2 \gg f_1$.

FM system

$$\text{FM noise power} = \int_0^{f_2} y^2\,\mathrm{d}f = \int_0^{f_2} \left(\frac{f}{f_3}\right)^2 \frac{\mathrm{d}f}{1+(f/f_1)^2} \qquad \left(y = \frac{f}{f_3}\right)$$

or $$\text{FM noise power} = \frac{1}{f_3^2}\int_0^{f_2} \frac{f^2\,\mathrm{d}f}{1+(f/f_1)^2}$$

Putting $f/f_1 = x$ or $f = xf_1$ with $\mathrm{d}f = f_1\,\mathrm{d}x$, we obtain

$$\text{FM noise power} = \frac{f_1^3}{f_3^2}\int_0^{f_2} \frac{x^2\,\mathrm{d}x}{x^2+1}$$

$$= \frac{f_1^3}{f_3^2}\left[\int_0^{f_2} \mathrm{d}x - \int_0^{f_2} \frac{\mathrm{d}x}{x^2+1}\right]$$

or $$\text{FM noise power} = \frac{f_1^3}{f_3^2}[f_2/f_1 - \tan^{-1} f_2/f_1] \simeq \frac{f_1^2 f_2}{f_3^2}$$

since $f_2/f_1 \gg \tan^{-1} f_2/f_1$.

Hence
$$\frac{\text{AM noise power}}{\text{FM noise power}} = \frac{(\pi/2)f_1 \times f_3^2}{f_1^2 f_2} = \frac{\pi \times f_3^2}{2 \times f_1 f_2}$$
$$= \frac{\pi \times (75 \times 10^3)^2}{2 \times 3{\cdot}1 \times 10^3 \times 15 \times 10^3}$$
$$= 190$$

or about 23 dB.

Since the same ratio *without* pre-emphasis and de-emphasis was found earlier to be 19 dB, the net improvement *with* the use of pre-emphasis and de-emphasis is about 4 dB.

Appendix D: Pulse modulation

The expressions for a PDM pulse train and a PPM pulse train will be derived assuming a single modulating signal $v = \sin \omega_m t$ and a sampling frequency $f_s = 1/T$.

Pulse duration modulation

The unmodulated pulse train is given by*

$$v_i(t) = \frac{\tau}{T} + \frac{2\tau}{T} \sum_{n=1}^{n=\infty} \frac{\sin(n\omega_s \tau/2)}{n\omega_s \tau/2} \cos n\omega_s t$$

or

$$v_i(t) = \frac{\tau}{T} + \sum_{n=1}^{n=\infty} \frac{2}{n\pi} \sin(n\omega_s \tau/2) \cos n\omega_s t$$

If the increase in pulse width due to the modulating signal is $\tau(1 + m \sin \omega_m t)$, where $m = \Delta\tau/\tau$, the modulated pulse train is given by

$$v_c(t) = \frac{\tau}{T}(1 + m \sin \omega_m t) + \sum_{n=1}^{n=\infty} \frac{2}{n\pi} \sin n\omega_s [\tau/2 + (m\tau/2) \sin \omega_m t] \cos n\omega_s t$$

$$= \frac{\tau}{T}(1 + m \sin \omega_m t) + \sum_{n=1}^{n=\infty} \frac{2}{n\pi} \cos n\omega_s t [\sin(n\omega_s \tau/2) \cos\{(mn\omega_s \tau/2) \sin \omega_m t\}$$
$$+ \cos(n\omega_s \tau/2) \sin\{(mn\omega_s \tau/2) \sin \omega_m t\}]$$

or

$$v_c(t) = \frac{\tau}{T}(1 + m \sin \omega_m t) + \frac{2}{\pi} \cos \omega_s t [\sin(\omega_s \tau/2) \cos\{(m\omega_s \tau/2) \sin \omega_m t\}$$
$$+ \cos(\omega_s \tau/2) \sin\{(m\omega_s \tau/2) \sin \omega_m t\}] + \ldots$$

* See F. R. Connor, *Signals*, Edward Arnold (1982).

Now

$$\cos\{(m\omega_s\tau/2)\sin\omega_m t\} = J_0(m\omega_s\tau/2) + 2J_2(m\omega_s\tau/2)\cos 2\omega_m t + \ldots$$
$$\sin\{(m\omega_s\tau/2)\sin\omega_m t\} = 2J_1(m\omega_s\tau/2)\sin\omega_m t + 2J_3(m\omega_s\tau/2)\sin 3\omega_m t + \ldots$$

Hence

$$v_c(t) = \frac{\tau}{T}(1 + m\sin\omega_m t)$$
$$+ \frac{2}{\pi}\cos\omega_s t[\sin(\omega_s\tau/2)\{J_0(m\omega_s\tau/2) + 2J_2(m\omega_s\tau/2)\cos 2\omega_m t + \ldots\}$$
$$+ \cos(\omega_s\tau/2)\{2J_1(m\omega_s\tau/2)\sin\omega_m t + 2J_3(m\omega_s\tau/2)\sin 3\omega_m t + \ldots\}]$$
$$+ \ldots$$

Since $\qquad 2\cos A\cos B = \cos(A+B) + \cos(A-B)$

and $\qquad 2\cos A\sin B = \sin(A+B) - \sin(A-B)$

hence

$$v_c(t) = \frac{\tau}{T} + \frac{m\tau}{T}\sin\omega_m t + \frac{2}{\pi}\sin(\omega_s\tau/2)J_0(m\omega_s\tau/2)\cos\omega_s t$$
$$+ \frac{2}{\pi}\cos(\omega_s\tau/2)J_1(m\omega_s\tau/2)\{\sin(\omega_s + \omega_m)t - \sin(\omega_s - \omega_m)t\}$$
$$+ \frac{2}{\pi}\sin(\omega_s\tau/2)J_2(m\omega_s\tau/2)\{\cos(\omega_s + 2\omega_m)t + \cos(\omega_s - 2\omega_m)t\}$$
$$+ \ldots$$

Since the Bessel functions $J_0(m\omega_s\tau/2)$, $J_1(m\omega_s\tau/2)$, etc. are similar in form to $J_0(\Delta\phi)$, $J_1(\Delta\phi)$, etc. for a PM wave, the expression for $v_c(t)$ represents a phase-modulated wave.

Pulse position modulation

Let the unmodulated pulse train be given by

$$v_i(t) = \tau f_s + \sum_{n=1}^{n=\infty} \frac{2}{n\pi}\sin(n\pi f_s\tau)\cos n\phi_s$$

where $f_s = 1/T$ and $\phi_s = \omega_s t$.

Due to the modulating signal, the centre of each pulse is shifted by an amount $\Delta t\sin\omega_m t$ or $mT\sin\omega_m t$, where $m = \Delta t/T$. Hence, the modulated pulse train can be considered as a case of *non-uniform* sampling, where the sampling frequency is now a function of time and is denoted by $f_s(t)$. The sampled phase $\phi_s(t)$ of any pulse position is given by

$$\phi_s(t) = \omega_s(t + \Delta t\sin\omega_m t)$$

with $$f_s(t) = \frac{1}{2\pi}\frac{d\phi_s(t)}{dt} = f_s(1 + m\omega_m T \cos \omega_m t)$$

where $\Delta t = mT$ is a constant for a given system.

The modulated pulse train is given by

$$v_c(t) = \tau f_s(1 + m\omega_m T \cos \omega_m t) + \sum_{n=1}^{n=\infty} \frac{2}{n\pi} \sin[n\pi f_s(t)\tau] \cos n\phi_s(t)$$

or

$$v_c(t) = \frac{\tau}{T} + m\omega_m \tau \cos \omega_m t$$
$$+ \sum_{n=1}^{n=\infty} \frac{2}{n\pi} \sin[(n\omega_s\tau/2) + (n\omega_s\tau/2)m\omega_m T \cos \omega_m t] \cos[n\omega_s t + mn\omega_s T \sin \omega_m t]$$

Hence $$v_c(t) \simeq \frac{\tau}{T} + m\omega_m \tau \cos \omega_m t$$
$$+ \frac{2\tau}{T} \sum_{n=1}^{n=\infty} \frac{\sin(n\omega_s\tau/2)}{(n\omega_s\tau/2)} (1 + m\omega_m T \cos \omega_m t) \cos n[\omega_s t + \phi(t)]$$

where $\phi(t) = m\omega_s T \sin \omega_m t$ and $(\omega_s \tau/2)$ is very small.

The first term is a d.c. term, while the second term varies at the modulating frequency. In addition to an amplitude factor, the third term represents a set of harmonics at the sampling frequency which are phase-modulated according to $\phi(t)$. If the side-frequencies of the first PM wave are sufficiently far from the modulating frequency, the modulation can be recovered by passing the modulated signal through a low-pass filter.

Appendix E: Random sequence spectrum[43]

The power spectrum of a random binary code sequence can be obtained by using the Wiener–Khintchine relationship between the power spectrum $S(\omega)$ and the autocorrelation function $R(\tau)$. They are given by

$$S(\omega) = \int_{-\infty}^{+\infty} R(\tau) e^{-j\omega\tau} d\tau$$

and $$R(\tau) = \lim_{T \to \infty} \frac{1}{2T} \int_{-T}^{+T} f(t) f(t - \tau) dt$$

where $S(\omega)$ is given in terms of the angular frequency ω and $R(\tau)$ is given in terms of the time displacement τ.

Assume that the random binary sequence shown in Fig. A.4 is made of bits of duration T seconds and with an amplitude $\pm A$. Also, assume the sequence is

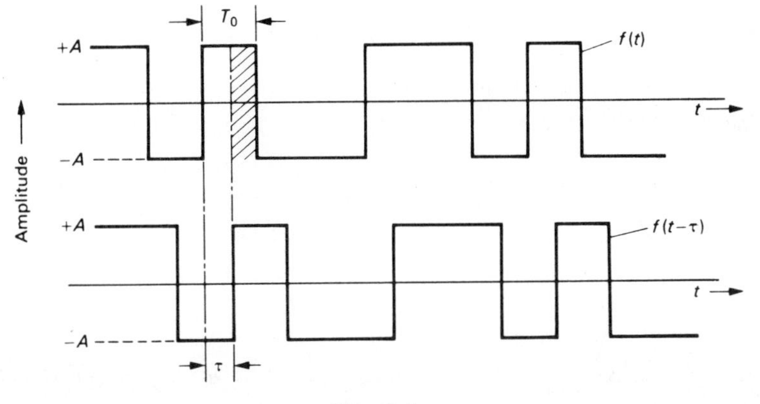

Fig. A.4

continuous with positive and negative excursions occurring with equal probability but in random fashion and such that successive pulses are independent.

For $|\tau| > T_0$, the average value of $f(t)f(t-\tau)$ is zero since its instantaneous value has an equal probability of being $+A^2$ or $-A^2$. For $|\tau| < T_0$, the average value over one pulse of $f(t)f(t-\tau)$ is given by the area of the overlapping part of the displaced pulses shown above which can be seen to have a value of $A^2(T_0 - |\tau|)$. Furthermore, as the equation for $R(\tau)$ is averaged over $2T$, and $2T/T_0$ pulses occur in this time, we have

$$R(\tau) = 0 \qquad \text{(for } |\tau| > T_0\text{)}$$

and

$$R(\tau) = \frac{2T}{T_0}\frac{A^2}{2T}(T_0 - |\tau|)$$

$$= \frac{A^2}{T_0}(T_0 - |\tau|) \qquad \text{(for } |\tau| < T_0\text{)}$$

which is shown in Fig. A.5(a).

The power spectrum is then given by

$$S(\omega) = 2\int_0^\infty R(\tau)\mathrm{e}^{-j\omega\tau}\,\mathrm{d}\tau$$

and substituting for $R(\tau)$ yields for positive frequencies only

$$S(\omega) = \frac{2A^2}{T_0}\int_0^{T_0}(T_0 - |\tau|)\cos\omega\tau\,\mathrm{d}\tau$$

$$= \frac{2A^2}{\omega^2 T_0}\left[1 - \cos\omega T_0\right]$$

or
$$S(\omega) = A^2 T_0 \frac{\sin^2 (\omega T_0/2)}{(\omega T_0/2)^2}$$

which is shown in Fig. A.5(b).

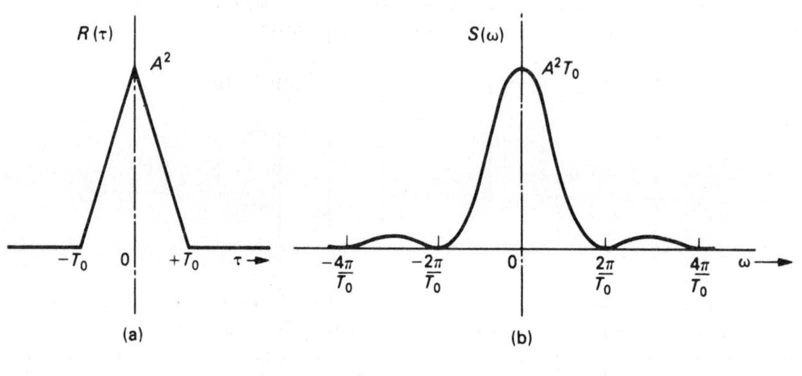

Fig. A.5

Appendix F: Radio receivers

Most radio receivers employ the superheterodyne principle in which the incoming signal is converted to a fixed, lower *intermediate frequency* or IF. The IF falls within the radio-frequency spectrum and hence the name superheterodyne receiver or *superhet.*

For broadcast reception, a single IF frequency is used but, for more sensitive communication reception, a double-superheterodyne receiver which uses two IF frequencies is generally employed. Various receivers are designed to receive either AM, or FM, or SSB signals. Though there are many features in common between these receivers, there are also some notable differences which distinguish one type of receiver from another.

AM receivers

A typical superhet receiver for AM broadcast purposes is shown in Fig. A.6. The RF stages are usually absent in the cheaper, domestic receiver but, in the more expensive receivers, one or two RF stages may be employed to give better sensitivity.

The incoming or tuned radio frequency is then mixed with the local oscillator frequency in the frequency-changer stage and the *difference* frequency or IF is selected in the output of this stage. Usually, the IF is around 455–465 kHz and, for practical convenience, the local oscillator frequency is above that of the incoming tuned frequency. As the IF is a fixed frequency, the tuned circuits are

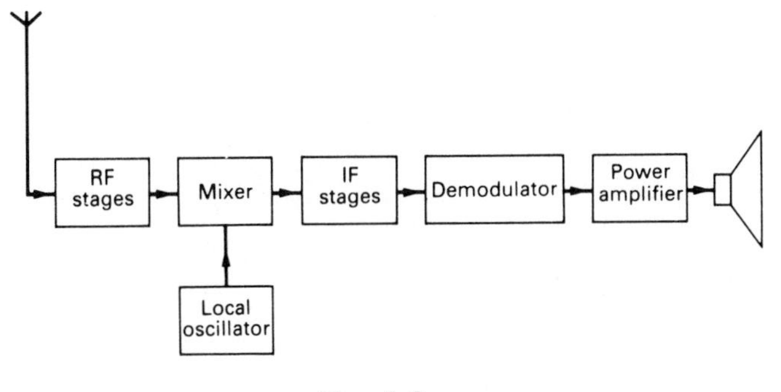

Fig. A.6

designed to follow or *track* one another as the receiver is tuned over a given frequency band.

The IF signal is subsequently amplified in two or three IF stages which are band-pass circuits designed to pass only the audio frequencies required for intelligible speech or commercial quality music. Hence, much of the receiver selectivity is obtained by these stages and also the gain required to provide the subsequent demodulation with a suitable drive signal for linear operation.

The demodulator is usually a linear diode detector which requires a fairly large input signal (several volts) to ensure linear operation and hence avoid serious harmonic distortion, especially in the case of music. The output from the detector is subsequently filtered and amplified by an audio-power amplifier in order to drive a suitable loudspeaker.

Various features, such as automatic gain control, volume control, and waveband selection, are provided in general purpose receivers.

For special communication purposes, a double-superheterodyne receiver is used employing a first IF of around 1·6 MHz to give good image channel rejection and a second IF frequency of around 100 kHz to give good adjacent channel rejection. Furthermore, one or more RF stages are invariably used, together with many useful features, e.g. automatic gain control, bandwidth selectivity, and noise limiter. In some special cases, the same receiver may be designed for AM or FM reception, in which case the first IF is usually 10·7 MHz while the second IF is about 465 kHz.

FM receivers

A typical FM broadcast receiver used for the VHF band is shown in Fig. A.7. The receiver usually tunes from 88 to 108 MHz and provides high-fidelity audio reception with a bandwidth of 15 kHz. Since the discriminator requires a much larger signal than does the AM detector, one or two RF stages are essential prior to the frequency changer. Typically, the IF is 10·7 MHz and two or three IF stages are used to provide the large drive signal for the discriminator.

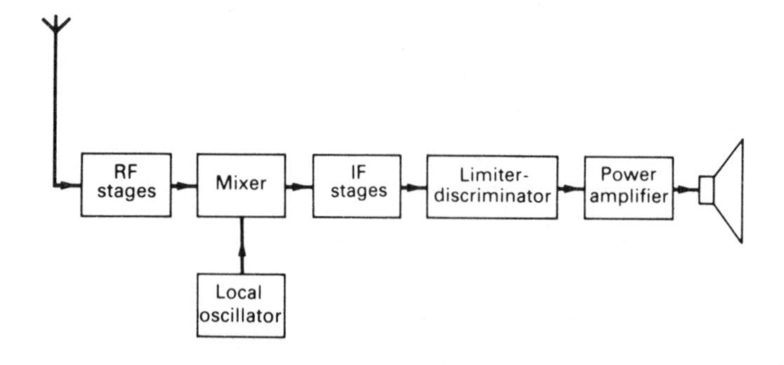

Fig. A.7

For good linearity, the Foster–Seeley discriminator is generally used and so it must be preceded by a limiter to be effective. In cheaper, domestic receivers the ratio detector is used as no separate limiter is required for its proper operation. The output audio signal is amplified by a suitable audio-power amplifier to drive the output loudspeaker.

In the double-superheterodyne communication receiver, the first IF is usually 10·7 MHz, as mentioned earlier, for good image channel rejection and the second IF is around 465 kHz to give good adjacent channel rejection. FM receivers can also be provided with the additional features that were mentioned earlier for AM receivers. In addition, since frequency stability is essential in FM, automatic frequency control is used within the receiver.

Since FM reception involves a threshold signal-to-noise ratio, special feedback receivers using phase-lock techniques may be employed to lower the threshold from about 10 dB to about 5 or 6 dB in usual cases. Further details will be found in Section 6.5.

Appendix G: Synchronous detection

The importance of phase or frequency coherence can be examined by assuming that the local injected carrier has a slightly different phase or frequency from its proper value. This will be done for both DSBSC and SSBSC signals using the circuit shown in Fig. A.8.

Phase coherence

Assume the local carrier frequency $\omega_c/2\pi$ is the same as that at the transmitter but there is an arbitrary phase difference ϕ. The output of the synchronous detector is $v_o = kv_i v_c$ and, for a DSBSC signal, we have from Section 6.1 $v_i = mV_c \sin \omega_c t \sin \omega_m t$ with $v_c = \sin(\omega_c t + \phi)$.

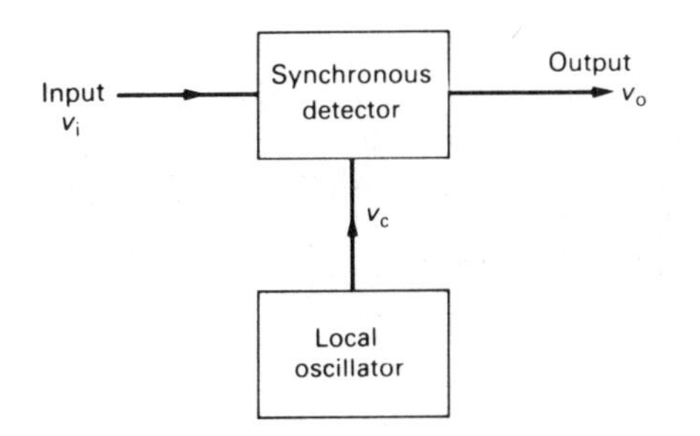

Fig. A.8

Hence
$$v_o = kmV_c \sin \omega_c t \sin \omega_m t \sin (\omega_c t + \phi)$$
$$= kmV_c \sin \omega_c t \sin \omega_m t \{\sin \omega_c t \cos \phi + \cos \omega_c t \sin \phi\}$$
$$= kmV_c \sin^2 \omega_c t \sin \omega_m t \cos \phi + \left(\frac{kmV_c}{2}\right) \sin 2\omega_c t \sin \phi$$
$$= \left(\frac{kmV_c}{2}\right)(1 - \cos 2\omega_c t) \sin \omega_m t \cos \phi + \ldots$$

or
$$v_o = \left(\frac{kmV_c}{2}\right) \sin \omega_m t \cos \phi + \ldots$$

The modulation is given essentially by the first term as the other terms are mainly harmonics of the carrier frequency. When $\phi = 0$, v_o is a maximum, but when $\phi = \pi/2$, $v_o = 0$ and the modulation is lost. Hence, as ϕ varies continuously, signal 'fading' will occur.

For an SSBSC signal, we have $v_i = (mV_c/2) \cos (\omega_c - \omega_m)t$ for a lower side-frequency only, hence
$$v_o = \left(\frac{kmV_c}{2}\right) \cos (\omega_c - \omega_m)t \sin (\omega_c t + \phi)$$
$$= \left(\frac{kmV_c}{2}\right) \sin (\omega_c t + \phi) \cos (\omega_c - \omega_m)t$$
$$= \left(\frac{kmV_c}{4}\right)[\sin\{(2\omega_c - \omega_m)t + \phi\} + \sin(\omega_m t + \phi)]$$

or
$$v_o = \left(\frac{kmV_c}{4}\right) \sin (\omega_m t + \phi) + \left(\frac{kmV_c}{4}\right) \sin\{(2\omega_c - \omega_m)t + \phi\}$$

The modulation is given by the first term and we observe that there is a phase delay due to ϕ, which is a maximum when $\phi = \pi$. However, for speech and music a small phase delay is not serious.

Frequency coherence

Since a frequency shift of $\delta\omega_c/2\pi$ is equivalent to a phase shift $\phi = \delta\omega_c t$, the effect of a frequency change can be evaluated by substituting $\delta\omega_c t$ for ϕ in the relevant expressions above. For an incoming DSBSC signal, the output v_o becomes

$$v_o = \left(\frac{kmV_c}{2}\right)\sin\omega_m t\cos\phi + \ldots = \left(\frac{kmV_c}{2}\right)\sin\omega_m t\cos\delta\omega_c t$$

or

$$v_o = \left(\frac{kmV_c}{4}\right)[\sin(\omega_m + \delta\omega_c)t + \sin(\omega_m - \delta\omega_c)t] .$$

and so the output contains two modulation frequencies which are slightly different from the original modulation, and this causes some distortion.

For an incoming SSBSC signal, the output v_o becomes

$$v_o = \left(\frac{kmV_c}{4}\right)\sin(\omega_m t + \phi) + \ldots = \left(\frac{kmV_c}{4}\right)\sin(\omega_m + \delta\omega_c)t$$

and here too there is some frequency distortion in the output signal, but due to one modulation component only.

Appendix H: Detectors

Square-law detector

The input signal with narrowband noise can be expressed by

$$v_i = V_c\sin\omega_c t + x(t)\sin\omega_c t + y(t)\cos\omega_c t$$

and the output current of the detector is given by $i_o = kv_i^2$, where k is a constant. Hence

$$i_o = k[\{V_c + x(t)\}\sin\omega_c t + y(t)\cos\omega_c t]^2$$

$$= \frac{k}{2}\Big[\{V_c + x(t)\}^2(1 - \cos 2\omega_c t)$$

$$+ \{V_c + x(t)\}\,y(t)\sin 2\omega_c t + y^2(t)(1 + \cos 2\omega_c t)\Big]$$

or

$$i_o = \frac{k}{2}\left[V_c^2 + 2V_c x(t) + x^2(t) + y^2(t)\right]$$

$$-\frac{k}{2}\left[V_c^2 + 2V_c x(t) + x^2(t) - y^2(t)\right]\cos 2\omega_c t$$

$$+\frac{k}{2}\left[V_c y(t) + x(t)y(t)\right]\sin 2\omega_c t$$

For a *large* input signal-to-noise ratio, i.e. $V_c \gg x(t)$, the output signal and noise powers in a $1\,\Omega$ load are

$$S_o = (kV_c^2/2) = k^2V_c^4/4$$
$$N_o = \overline{[kV_c x(t)]^2} = k^2V_c^2\overline{x^2(t)}$$

as the noise contributions from the terms $x^2(t)$ and $y^2(t)$ may be neglected. Similarly, for the input signal and noise powers, we have

$$S_i = (V_c/\sqrt{2})^2 = V_c^2/2$$
$$N_i = \tfrac{1}{2}\overline{[x^2(t) + y^2(t)]} = \overline{x^2(t)}$$

as $x(t)$ and $y(t)$ are independent Gaussian variables. Hence

$$S_i/N_i = V_c^2/2\overline{x^2(t)}$$

and

$$S_o/N_o = V_c^2/4\overline{x^2(t)}$$

or

$$(S_o/N_o) = \tfrac{1}{2}(S_i/N_i)$$

and so there is a 3 dB loss through the detector.

Comment
More generally, it can be shown that

$$\frac{S_o}{N_o} = \frac{(S_i/N_i)^2}{1 + 2(S_i/N_i)}$$

and for very low values of S_i/N_i we obtain

$$(S_o/N_o) \simeq (S_i/N_i)^2$$

Linear detector
The input signal with narrowband Gaussian noise can be expressed by

$$v_i = V_c \sin \omega_c t + x(t) \sin \omega_c t + y(t) \cos \omega_c t$$

or

$$v_i = A(t) \sin (\omega_c t + \phi)$$

where $A(t) = [\{V_c + x(t)\}^2 + y^2(t)]^{1/2}$ and $\phi = \tan^{-1}\left[\dfrac{y(t)}{V_c + x(t)}\right]$

The output voltage v_o is the peak value, which is the envelope of v_i, and so

$$v_o = A(t) = [\{V_c + x(t)\}^2 + y^2(t)]^{1/2}$$

For a *large* input signal-to-noise ratio, i.e. $V_c \gg x(t)$, the output signal and noise powers in a $1\,\Omega$ load are respectively

$$S_o = V_c^2$$
$$N_o = \tfrac{1}{2}\overline{[x^2(t) + y^2(t)]} = \overline{x^2(t)}$$

as $x(t)$ and $y(t)$ are independent Gaussian variables. Hence

$$S_o/N_o = V_c^2/\overline{x^2(t)}$$

As before, the input signal and noise powers are respectively

$$S_i = V_c^2/2$$

and
$$N_i = \tfrac{1}{2}\overline{[x^2(t) + y^2(t)]} = \overline{x^2(t)}$$

with
$$S_i/N_i = V_c^2/2\overline{x^2(t)}$$

Hence
$$(S_o/N_o) = 2(S_i/N_i)$$

and so there is a 3 dB gain through the detector.

Comment
For low values of S_i/N_i, the linear detector behaves as a square-law detector and so we have

$$(S_o/N_o) \simeq (S_i/N_i)^2$$

Synchronous detector
If the input signal with narrowband Gaussian noise is v_1 and the coherent local oscillator signal is v_2, we have

$$v_1 = V_1 \sin \omega_c t + x(t) \sin \omega_c t + y(t) \cos \omega_c t$$
$$v_2 = V_2 \sin \omega_c t$$

and the output voltage is $v_o = kv_1v_2$, where k is a constant of proportionality. Hence

$$v_o = k\{V_1 + x(t)\} V_2 \sin^2 \omega_c t + kV_2\, y(t) \sin \omega_c t \cos \omega_c t$$

or
$$v_o = \frac{kV_2}{2}\{V_1 + x(t)\}(1 - \cos 2\omega_c t) + \frac{V_2}{2}\, y(t) \sin 2\omega_c t$$

The output signal power in a 1 Ω load is therefore

$$S_o = k(V_1 V_2/2)^2$$

and the output noise power *coherent* with the signal is given by

$$N_o = k(V_2/2)^2\overline{x^2(t)}$$

Hence
$$S_o/N_o = V_1^2/\overline{x^2(t)}$$

As before, the input signal and noise powers are respectively

$$S_i = V_1^2/2$$

and
$$N_i = \overline{x^2(t)}$$

with
$$S_i/N_i = V_1^2/2\overline{x^2(t)}$$

Hence $$(S_o/N_o) = 2(S_i/N_i)$$

and so there is a 3 dB gain through the detector for all values of S_i/N_i.

Appendix I: Feedback loops

Frequency-locked loop

A schematic diagram of the frequency-locked loop is shown in Fig. 6.11. Neglecting delays in the loop and assuming a noiseless input signal $v_i(t)$, we have

$$v_i(t) = V_c \sin[\omega_c t + \theta_i(t)]$$

where $\theta_i(t)$ is the instantaneous phase angle which depends on the type of signal modulation used. Assuming that the angular frequency changes by a peak amount $\Delta\omega$ per unit modulating voltage $m(t)$, then

$$\theta_i(t) = \Delta\omega\, m(t)$$

and the instantaneous signal frequency ω_i is given by

$$\omega_i = \omega_c + \frac{d}{dt}[\theta_i(t)] = \omega_c + \Delta\omega m'(t) = \omega_c + \dot{\theta}_i(t)$$

since $\dot{\theta}_i(t) = \Delta\omega m'(t)$.

If $v_d(t)$ is the output voltage of the discriminator, let $\beta v_d(t)$ be the change in the VCO output due to the feedback factor β. The corresponding change in the difference frequency signal is $\Delta\omega\, m'(t) - \beta v_d(t)$ and, for linear operation, the discriminator output is given by

$$v_d(t) = K_d[\Delta\omega\, m'(t) - \beta\, v_d(t)]$$

where K_d is the discriminator constant in volts/Hz. Hence, we have

$$v_d(t) = \frac{K_d}{1 + \beta K_d} \Delta\omega m'(t)$$

and $$v_o(t) \simeq A\dot{\theta}_i(t)$$

if $\beta \gg 1$, $A = 1/\beta$, and $\dot{\theta}_i(t) = \Delta\omega\, m'(t)$.

This result shows that an input change in angular frequency $\Delta\omega$ is effectively reduced by the loop feedback factor β and, consequently, the band-pass filter bandwidth can be much narrower than the signal excursion bandwidth if β is large. It is usually designed to be wide enough to accept one pair of side-bands.

Phase-locked loop

A schematic diagram of the phase-locked loop is shown in Fig. 6.12. Assuming the input and feedback frequencies are the same, the phase comparator detects the phase difference between the input and feedback voltages. For a sinusoidal input, there must be a 90° phase difference between these voltages and we have

$$v_i(t) = A_1 \sin[\omega_c t + \theta_i(t)]$$

$$v_f(t) = A_2 \cos[\omega_c t + \theta_f(t)]$$

and
$$v_c(t) = KA_1 \sin[\omega_c t + \theta_i(t)] A_2 \cos[\omega_c t + \theta_f(t)]$$

After low-pass filtering, this yields

$$v_c(t) = K_c \sin[\theta_i(t) - \theta_f(t)]$$

where $K_c = KA_1A_2/2$ and, if $\theta_i(t)$ and $\theta_f(t)$ are small, we obtain linear operation. Hence

$$v_o(t) \simeq K_c[\theta_i(t) - \theta_f(t)]$$

Since the output of the loop filter is $v_o(t)$, the VCO generates an output voltage whose angular frequency is given by $\omega_f(t) = K_f v_o(t)$, where K_f is a constant and its arbitrary phase $\theta_f(t)$ is expressed by

$$K_f \int_0^t v_o(t) \mathrm{d}t = \theta_f(t)$$

or
$$v_o(t) = \frac{\mathrm{d}}{\mathrm{d}t}\left[\frac{1}{K_f}\theta_f(t)\right] = \frac{1}{K_f}\dot{\theta}_f(t)$$

For high loop gain, $\theta_i(t) \simeq \theta_f(t)$ and so we obtain

$$v_o(t) \simeq A\dot{\theta}_i(t)$$

where $A = 1/K_f$ and the result shows that the phase-locked loop behaves as an FM demodulator.

Index